The Rooftop Beekeeper

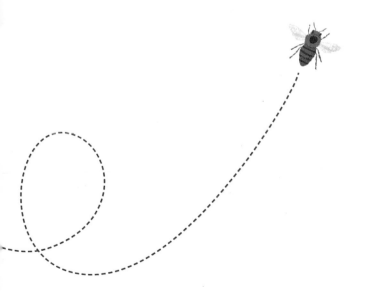

The Rooftop Beekeeper

A SCRAPPY GUIDE TO KEEPING URBAN HONEYBEES

Megan Paska

With
Rachel Wharton

Photographs by
Alex Brown

Illustrations by
Masako Kubo

CHRONICLE BOOKS
SAN FRANCISCO

Library of Congress Cataloging-in-Publication Data available.
ISBN 978-1-4521-0758-5

Manufactured in China

Designed by Suzanne LaGasa
Illustrations by Masako Kubo
The photographs on pages 17, 46, 52, 55, 83, 85, 95, and 124 were shot on location at Brooklyn Grange.

10 9 8 7 6 5 4 3 2 1

Chronicle Books LLC
680 Second Street
San Francisco, California 94107
www.chroniclebooks.com

Contents

How a City Girl Got Stung

Sometimes when I stop to think about the fact that I am an urban apiarist—meaning people pay me to take care of bees—it still seems a little crazy.

For one, I live in New York City, in its most populous borough, Brooklyn. It's noisy, it's grimy, and, at first glance, it seems like everything is covered in either pavement or an apartment tower. The concrete jungle is not exactly the backdrop one imagines as a honeybee heaven. When I got my first batch of bees on a blustery Easter morning—the buzzing package headed for my Brooklyn rooftop—I was skeptical that my new hobby made any sense. Would my bees starve? Would they get sick? Would they annoy my neighbors?

But the real reason my current occupation is so strange is that for most of my life, I was afraid of bees. Terrified. When I think back to my first experience with the tiny, mighty European honeybee—a.k.a. *Apis mellifera*—I can say with certainty that it was not a "Eureka!" moment. Or at least it was not a good one. I was about six, and I was playing barefoot in a weedy field of clover and dandelion behind the apartment complex in Baltimore where I lived. I was jolted by three sudden pinpricks, which evolved into a searing, pulsating pain that engulfed my entire foot.

I ran home, crying and confused, and burst through my front door, red faced and hysterical. I didn't know what could have caused that sort of pain, but it was certainly something devilish. But my Grandma Dorothy calmly sat me down and pointed out the stingers, which I know so well today, still lodged in the top of my foot with the telltale venom sac of a honeybee still attached. With the blunt edge of a kitchen knife, she knowingly scraped them out and anointed my plump, red foot with baking soda paste. ("Scrape; never pinch," she told me. It's the cardinal rule in removing a bee stinger, or else you'll force more venom into your bloodstream.) In the end, my dramatic first meeting with the honeybee ended in three fatalities—the bees—and a tender, fat foot. Grandma told me that it would hurt for a while, and she was right: It was the most traumatizing thing I had experienced thus far in my short life, and that day I swore that I would never again get anywhere near those ornery things as long as I lived.

But, two decades later, there I was in Brooklyn, ordering bees from a fledgling urban bee club and building a hive on the kitchen floor of my apartment. By that point, I'd discovered that urban beekeepers were multiplying in France and the United Kingdom. And I'd learned that bees in the city have just as good a chance as anywhere else to thrive. Urban trees and overgrown lots provide enough nectar and pollen from weeds like yellow sweet clover, curled blooms of gooseneck loostrife, or yellow tufts of goldenrod not just to sustain my bees throughout the seasons but to score me some of their surplus honey.

My neighbors hardly notice the busy apiary situated on the rooftop above our heads and, in fact, are reminded of its presence there only when they are gifted small jars of bee goodness in the summertime. That's Brooklyn honey, and, in case you're wondering, it is really good!

I get asked a lot: "How did you decide to become a beekeeper?" Truth be told, it took a few decades. Things started to change when I got a little older and began spending more time visiting my family's farm, a 450-acre parcel outside of Lynchburg, Virginia. Until I was a teenager, my mom, baby sister, and I would go to the farm every summer when school let out. It was always a total alien experience; I was a Baltimore city kid through and through, but this vast, quiet place with its verdure and its kindly people grew on me. I began to think of it as a second home.

My family in Virginia had held on to a tradition that us city kin had long since abandoned. They grew and raised a significant amount of what they cooked, and they really seemed to savor it when it came time to eat it. It didn't hurt that my Aunt Joanne was a great country cook: Most of the savory things she made were flavored with salty bacon fat rendered from the pigs the family had raised and sent off to be butchered and cured nearby. Jams and jellies were made right there in her kitchen, put up in jars, and kept in the cellar under her house until they were ready to eat. I had my first vine-ripe, homegrown tomato on an early trip to Virginia. It was shocking to me—a kid who hated tomatoes that weren't in the form of spaghetti sauce or ketchup. They were wonderfully flavorful and juicy and not at all mealy and bland like the ones topping the fast-food burgers I was used to. But at Aunt Joanne's house, you could find them on the kitchen table at each meal, sliced and lightly salted. I'd eat them sandwiched between two pieces of fresh baked bread with a little smear of smoky bacon grease. I never looked at a tomato the same way again. But even more important, the farm taught me to understand where my tomato came from and to realize that I was capable of making it, not just buying it.

A few years later, I signed the lease on my first house, a rustic limestone structure built for the sailmakers that worked in factories along the Jones Falls in Maryland. The house dated back to the 1840s, and it had three fireplaces and wooden plank floors. It was wonderfully old and dank. You could feel the spirits of generations of transplanted Appalachian mill workers emanating from the worn wood floors and plaster walls. But what I reveled in most was the sunny private backyard and the opportunity to grow food for real. I wanted more than just a couple of tomatoes and some herbs in containers; I wanted to stop buying substandard produce from supermarkets, save money, and be more like my Aunt Joanne. So I grew okra, lettuce, peas, peppers, summer and winter squash, tomatoes, and herbs in varieties that I had never seen or tasted before. I was surprised at how successful it was and how at ease I felt working outdoors.

Once I began harvesting vegetables, I had to figure out what to do with them. This was challenging: I still felt like an inexperienced cook, and after spending weeks coddling my kale as it was growing in the ground, I didn't want to ruin it. So I started relying on recipe books that my mother always had around but never really used. *The Joy of Cooking, Fannie Farmer Cookbook* . . . you know, old-school cookbooks. I'd spent time as a child lying on the floor dreamily flipping through pages of these like they were teen magazines, and now I was finally putting them to use. And once I got comfortable with my old-school kitchen skills, I started to wing it.

What does all that have to do with bees? In the garden, I'd started to notice that certain crops, like my butternuts, had different flowers. Male flowers with longer stalks didn't end up producing fruit, but the female flowers closer to the vine would wither and a small bulb—the start of a squash!—would appear in their place. But, I noticed, this would only happen if an insect, maybe a bee, had visited a male flower first, coating themselves with the vibrant dust we call pollen before inadvertently transferring it to a female blossom. I started to pay attention to similar crops, the ones that rely on pollination to produce: tomatoes, peppers, eggplant, berries, cucumbers, melons. . . . As a garden-obsessed adult, I'd finally accepted that bees weren't actually vicious at all, largely because I hadn't been stung since that first time. The bees kept to themselves, resolutely bound to their eternal task of finding food. I stayed out of their way; they stayed out of mine. As a result, I got over my bee anxiety, and I could appreciate them from a distance. And I began to wonder, where have all the honeybees gone? Why am I not seeing as many as I remember running away from as a kid?

I wouldn't have an answer to that complicated question for years. Especially because my next hobby wasn't bees, but home brewing. But life, as it so happens, has a funny way of sometimes bringing the right path to you even if you are too oblivious to head down it on your own. As it happens, I was drinking homemade beer and eating cheap pizza with some other nerds at a home brew meet-up in the back of a liquor store in the boondocks, and I met a beekeeper who made his own mead, or honey wine, from the liquid gold harvested from his very own apiary. That's really cool, I thought. And then I sampled it, reluctantly. It wasn't the saccharine swill I was expecting based on an earlier sample in high school, when I jokingly referenced *Beowulf* while choking down the overly sweet, astringent liquid. This man's mead was a different kind of beast altogether. It was made from the cured nectar of bees who fed from the tulip poplar tree. It was sweet, yes, but it was also berrylike, even juicy. I had to know more.

I picked this beekeeper's brain over those bad pepperoni and cheese slices, starting with mead but quickly turning to his second craft—managing stacked boxes of those flying, venomous insects that still freaked me out. I had never understood just how interesting bees were until he started dropping knowledge for me. Finally, visibly agitated with my endless barrage of questions,

he suggested that I sign up for a short course on beekeeping offered at the nature center just down the road from where we were drinking beer.

That winter I enrolled. It was a four-session, twelve-hour beginner's course—one night a week—taught by the state apiary inspector. I learned the ins and outs of keeping a beehive; honeybee anatomy; procuring a "package," the name for a complete set of bees that arrives at your doorstep; and, the most exciting part to me at the time, harvesting the honey. Even though I wondered how I would feel about them when I was finally stung again—which is inevitable for a beekeeper—in the spring, I decided I would keep some bees of my own upon the stone wall that surrounded my garden. I could see them floating from bloom to bloom, pollinating my vegetable garden and doubling my harvests. But just when I began to game plan, I was offered a job in New York City. It was unexpected but, like most people, I jumped at the chance to try a different life on for size. I traded in my shovel and trowel for a MetroCard and a messenger bag and headed off to the big city.

My transition to living in an even bigger city went pretty smoothly, at least at first. I had a good job working for a small company in a trendy part of Manhattan. I landed a great apartment in an up-and-coming Brooklyn neighborhood. I made enough money that I could drink over-priced cocktails and artisan-crafted beer at hipster bars and eat at all of my favorite restaurants every night of the week. By all accounts I was living the life, but some kind of void left me staring at my ceiling in bed each night wondering what in the hell I was doing with myself. Work, spend, sleep. Work, spend, sleep. It was a routine that really clashed with my conscience. In the largest, most populous city in the country, I managed to feel disconnected from the world.

In an attempt to fix what felt broken in my life, I ended up doing what most sensible people do in my position: I thought back to the time when I was the most fulfilled and hopeful. That was when I was working in the garden and growing my own food; when I had a real sense of home. Intensive gardening was a strange passion for a young person living in a city that never sleeps, but it was what I needed and, by God, I was going to have it again if it was the last thing I did.

Luckily things just sort of fell into place. The building I lived in was sold to two young women who promptly tore down the shoddy above-ground pool that had claimed our backyard. A plan coalesced: We were going to build raised beds for a garden. And they offered those of us in the building rooftop access, so we could go up and enjoy our view of midtown Manhattan across the East River. It was like the cosmos had heard my wishes. But the most serendipitous moment came a few weeks after we first planted our city crops. I was standing in line at a health food store during my lunch break, and I read an article about New Yorkers breaking the law and keeping bees. I burst through the door back at my office, beaming at my co-workers who had spent the last few

months listening to me go on and on about my Brooklyn garden. "Look!" I said, waving the magazine. "You guys won't believe this! I've found my people. Beekeepers here, of all places!"

I was elated, but I was still skeptical I'd ever be able to join their ranks. You've got to be either terribly lucky or terribly sneaky to put hives on your roof without anyone knowing or caring, I thought. I couldn't imagine deceiving anyone and getting away with it, so I was up-front with my landlords about my plan. I was blown away when they took very little convincing, even given beekeeping's illegality at the time. "We like honey," they told me. "If you say it's safe, let's do it! Put 'em on the roof." Even my neighbors didn't seem to mind. Most of them were Polish immigrants, who were used to beekeeping.

I promptly started going to meetings of a local beekeeping group. People from all walks of life would come to our monthly gatherings. Some had bees, but many just wanted them. Potential apiarists would show up with questions; those who already kept bees had stories to share. It was a comfort to find myself in the company of people who didn't find the need to have a little piece of the country in the city all that strange. A group of us got together and put in our orders for hive equipment, tools, and bees. One by one, we helped each other assemble our hives as they arrived in the form of unpainted, stacked planks of pine, all wrapped tightly with plastic.

I constructed mine with help from my boyfriend, both of us assigned the tedious task of gluing and tacking frames together on my kitchen floor. I set up the hive in my living room and imagined what it would look like with bees in it, little foragers zooming from their home, out into the world and back. I longed for my bees, and I didn't even have them yet. This was quite a change from the bee-fearing young woman I'd been just a few years before. In fact, when my first package of bees was delivered to my doorstep by a beekeeping friend one blustery Easter morning—within it a pulsing, buzzing mass of wings and venom—I was shaking like a leaf from excitement. I hadn't slept a wink; I couldn't recall being that rattled by anticipation before. I gingerly carried the box up through the hatch in my roof to the empty shell that would soon be the bees' home. The rising sun illuminated the face of the hive—"bring them to me," it seemed to smile—and I felt the crackle of something magical. What I was about to do felt right.

I sprayed down the ventilated part of the box of bees with sugar syrup as I'd been taught, my hands pink and numb from early April's chill. I took my shiny, still-virgin hive tool, and pried off the thin sheet of plywood covering the opening. I removed the queen cage and the can of syrup that had been sustaining my bees while they were in transit, and then I inverted the whole thing over my open hive. With a hearty downward swoosh, I officially began life as a beekeeper.

I've seen a lot and learned even more since that morning. I've expanded my own home apiary to three hives. I've started managing hives for other people, too: restaurants, urban farms, and even primetime TV shows. I've harvested massive quantities of honey and sold them to support my hobby. I've seen bees survive the winter to beget new colonies and new queens. I've cleaned out dead hives, misty-eyed at my failure to keep them alive. I've shaken swarms out of trees only to have them fly off to live in some fallen tree or brick wall. Every mistake, success, heartbreak, and victory has me more in love with the honeybee than I ever imagined I could be. I can't even look at a photograph of a bee without welling up with affection.

Luckily my love for bees has not gone unnoticed. I've had the opportunity to teach introductory beekeeping classes in New York City to new legions of wanna-bees, and now I've been given this chance to share what I've learned with you and many others. Hopefully I can not only pass on some knowledge but also plant a seed of passion and affection for the magical creatures that do so much for us and ask for nothing in return except, well, to "just bee."

That's why in this book I focus on "minimally invasive" hive management practices. I believe that bees know more about how to be bees than we do: To my mind, facilitating their long-term survival takes precedence over increasing their usefulness as pollinators or producers of a high-value commodity like honey. This is not the way everyone keeps bees, but plenty of information about other, more-involved methods of hive maintenance already exist out there. (Some of them are listed in the Resources, page 168, in case you want to decide for yourself.) It is my opinion that the other end results of beekeeping—honey and pollination—will come on their own if your bees are hardy and resilient. It's my goal to persuade you to grow strong bees, as many of them are not so strong these days. Honeybees are getting pummeled from all sides, from pesticides, new diseases, stress from being transported great distances for commercial pollination, a lack of adequate nutrition in their food sources, or even a lack of genetic diversity. As hobbyists, it truly is within our power to help them.

Still, if you're worried that you can't keep bees healthy because you don't live in the country, as you'll read in the very first chapter, cities can actually be some of the best places to keep a few hives. Unlike keepers living in rural towns, we city dwellers don't have to worry about pesticides from conventional farms spraying their fields. Rooftop hives also get ample sun and dry out faster after heavy rains; the ability to more-easily regulate temperature and humidity means bees with fewer diseases. But more important, at least from my point of view, urban apiaries give city dwellers an opportunity to commune with the natural world in a small but very profound way.

In fact, in chapter 1, I talk about another community—the other beekeepers in your region—and the importance of staying in touch with this group. If you do decide to keep bees, don't close yourself off. If you were to comb through this book and every other book on the market, rest assured you'd still have questions and real obstacles to overcome. That's when to turn to other beekeepers; you will find many who will have your back when something happens that you feel you can't handle alone. But it's also to every beekeeper's benefit to adapt the advice you are given; you have to decide what works best in the context of your own lifestyle and your own bees. The bees will tell you what they need, if you are listening. In the meantime, just remember that all beekeepers make mistakes; the best of us learn from them. Don't let the fear of failure stand in your way, ever.

It's my hope that as you read this book—learning about bee anatomy, colony management, or honey collection—you'll grow confident enough to plan your own urban apiary. Be fearless; simply do it. This book is meant to be a primer for making it happen. In fact, it follows my own decades-long path to becoming a beekeeper—from daydreaming to reading to doing. So get ready to score yourself a smoker, a veil, and a hive tool—and, even more important, your very own honeybees. Just be prepared; you might fall in love with being a beekeeper when you least expect it.

> "You will do foolish things, but do them with enthusiasm."
>
> – *Sidonie-Gabrielle Colette*

CHAPTER 1:

Why Keep Bees? Here's Why You Should Consider Becoming an Urban Apiarist

When I teach Urban Beekeeping 101, I start the class the same way every time: I list the full spectrum of good things honeybees have to offer. Why start with the positives instead of all the real strategic considerations of running an urban apiary? Sitting nervously in front of me are at least a dozen people who want to keep bees, and I want them to do it! This chapter, in essence, is my sales pitch, and it mirrors my classroom presentations. If any of the following strikes a chord, then you should seriously consider becoming a beekeeper.

THE COMMUNITY

Cities can be a tough place to practice self-reliance. When I first mentioned my desire to keep bees on my rooftop to my older Brooklyn friends, I got a fair share of head shaking if not outright dismissal. I had already been growing vegetables in raised beds in my 800-square-foot backyard; that was something they all could appreciate, especially when boxes of sweet little currant tomatoes accompanied my visits to their apartments. But bees? At the time, these so-called friends seemed to think I was losing touch with normalcy: "You should just move to the country if you want to do stuff like this," they told me. "It's crazy."

Beekeeping has long held a reputation for being a pretty solitary hobby, a pastime meant for the backwater. When most people close their eyes and conjure up somebody who keeps bees, they no doubt visualize some rickety white-haired geezer way out in the country wearing a pair of dirty overalls . . . or some extreme apiarist wearing a beard of bees. And, until now, that maybe wasn't too far from the truth: Currently, the average age of career beekeepers in America is around sixty.

But that's all changing, especially in cities where an ever-growing crop of young people are becoming ever more enthusiastic about supporting the local-food movement. We're all becoming more aware of how our food is made and at what price, ethically and environmentally. We are also beginning to command a better product as a response. And a lot of us want to know our daily decisions at our corner store do more than merely minimize damage to the environment: We want to feel like we are improving the world in some way.

The next generation of hobbyists often falls well under the age of forty, and the people you'll see enrolled in beekeeping, gardening, or composting classes are a diverse and modern lot concerned not with keeping the modern world out but with what sort of world their children will inherit. Many beekeepers see our hobby as a way to make a positive contribution to our immediate environment, especially given the other agrarian limitations of city life. And why can't we have the best of both worlds, meaning access to the rich culture and diversity of urban life as well as to homegrown food and relative self-sufficiency?

It turns out that you can. I found loads of folks who would become my friends and allies in less than a year—people who had started a rooftop-farm education program, others who were running a collective of community farming sites, another group turning empty school-yards into productive plots or shady community gardens into chicken coops where eggs were going for double the supermarket price per dozen. When the rest of my twenty-something peers thought I was out of my gourd, a small, but quickly growing, group of urban farmers were there to cheer me on, lend a helping hand, and then afterward, share dinner or have a few beers.

Plus, as a beekeeper, I was a vital link in

their chain of urban gardens and farms. My team of fuzzy pollinators would help to increase the yields on all fruit- and seed-producing crops growing within three miles (or more!) of my hives. Thanks to my bees, pepper blooms were pollinated as fast as they opened, while squash plants grew heavy with fruits later in the summer. Herbs flowered and went to seed, allowing frugal backyard gardeners to save them for next year's gardens. But you probably already knew from grade school that bees play an undeniably critical role in the cycle of plant life and reproduction. What you may not have expected is that they will also play a role in your connection to the people around you.

And that's just the urban agriculture scene. Another community I connected with was that of foodies and other culinary creative-types. Local chefs, retailers, and other food sellers are some of the first customers an urban beekeeper will have. The honey from my Brooklyn bees was a highly

Bee clubs are an excellent place to get your hands in a hive before deciding to commit to a colony of your own.

BROOKLYN'S
URBAN FARMING COMMUNITY

Where I live, you'll find a burgeoning community of like-minded folks working their fingers to the bone to prove that you can grow wholesome, fresh food in backyards, unused lots, and neighborhoods where there typically is little or no land. And there likely is a similar group of people where you live, too. Some do it for fun, others to feed and challenge themselves, and some to empower their neighbors and build strong communities. Still others do it to improve public health in food deserts, those neighborhoods where corner stores, gas stations, and delis are the only businesses selling food. As a local beekeeper, I am lucky enough to get to work with almost all of these people.

sought-after commodity, thanks to its rarity and its distinctive flavor. In my first season harvesting honey, I scored invites to set up tastings at local artisan food markets where I met an amazing array of enthusiastic chefs and entrepreneurs. I got offers to have my honey served at well-respected restaurants. Chefs and business owners started asking me to help them with their own rooftop bees so they could offer house honey on their menus. From there, I started teaching classes for those who wanted to learn more. In fact, just mention that you might keep bees or produce local honey to someone with a culinary background and watch them light up and demand to know more.

And they would be right to want to know. Everything about the process of honey making is not just intriguing but also really cool. From the way the bees

forage to the way we harvest the honey, it's all magical to almost anybody with taste buds. The best part of the experience is the sense of time and place—some people call that *terroir*—that you get when you taste your homegrown product for the first time. Contemplating bees is fun for a few, but honey is loved by nearly everyone. Hand a stranger a lovely jar of honeycomb or a piece of good bread dripping with honey that you've harvested yourself, and you can count on not just a smile but also a new friend.

One "agtivist" in my community is Stacey Murphy, founder of BK Farmyards, who took the idea of the community garden a step further than the rest: to the backyard. "New York City has more than 52,000 acres of backyards," says Stacey. "That's a lot of space to grow healthful food! BK Farmyards gives these existing green spaces the additional use of food production. As a network, these farmyards are reproducible, cost effective, and beautiful." In addition to establishing decentralized garden projects throughout Brooklyn's underutilized backyards and schoolyards, Stacey has also created a fifty-hen egg community-supported agriculture (CSA) project and a beekeeping project, as pollination is such a critical element for urban growing. The apiaries that Stacey establishes near her minifarms help ensure that the farms produce an abundant crop of vegetables. Good harvests not only will help feed the community each season but also will provide seeds for lower-cost seasons to come.

While Stacey goes to backyards, other Brooklyn farmers I've helped take their operations to the skies. Annie Novak of Eagle Street Rooftop Farm, for instance, is the co-founder and steward of a 6,000-square-foot green roof that produces an array of vegetables, herbs, and flowers for local restaurants and residents. The farm also hosts three apiaries, a small flock of egg-laying hens, and a nest of rabbits. Utilizing elements of permaculture, she's created a mini-ecosystem where one would otherwise not exist. In this system, the livestock devour vegetative debris and turn it into valuable, nutrient-rich compost, while the bees I helped to install assist in pollinating Annie's market crops.

Some farms continue to push the envelope. Take the thirty-hive urban bee yard operated by my friends at another massive rooftop farm called Brooklyn Grange. This apiary will not only serve as a way to produce honey on a much bigger scale for Brooklyn, but it already functions as a place to educate new beekeepers and raise even more bees adapted to urban life.

REASON #2:
POLLINATION ACROSS THE NATION

Pollinators like honeybees or the bumblebee here are essential in a plant's ability to self-propagate.

As wonderful as it is to make new friends or develop a sense of community, these are probably not reasons enough to keep bees. But how about adding "feeding the planet" to that list? Bees perform a duty that far surpasses that of a popularity magnet or even of making sweet honey: The primary and most beneficial function of the honeybee is pollination. For plants to produce fruit and seeds—which include the fruits and vegetables we eat—they have to be pollinated.

Pollination is what happens when a few grains of protein-rich pollen (scientists call those "gametes") go from the stamen, or male flower part, to the stigma, or female flower part. What happens next? The flower bud swells into a tomato or chile or cucumber or peach or strawberry—a recognizable fruit or a seedpod, in other words. Not only do we get to eat them, but also the plant can multiply or propagate itself for future seasons through those seeds. This is one of the most vital components of plant reproduction, and many plants can't do it without the assistance of a pollinator. Of course, honeybees are not the only natural pollinators out there; other valuable pollinators include wind, flies, birds, butterflies, hornets, and wasps, as well as different species of bees. But honeybees are special because each bee colony has a ton of them, making them a boon to any plants needing pollination within a few miles or kilometres of their home. The honeybee is the most efficient pollinator on the planet.

In fact, one of every three bites of food we take is the direct result of insect pollination. It's an idea you might have heard before—especially with recent articles about the plight of the honeybee—but it bears repeating. Orchard fruits, berries, nuts, a wide array of vegetables, and even cotton require the instinctive behavior of bees to help them pass forth their genetics into the future. Real farmers get a significant increase in yield when commercial bee pollination enters the equation, and the increase

in those yields can mean the difference between a profitable farm and one that fails. Without bees, we consumers truly can expect to see higher prices for food and possibly even shortages of food, which, over time, can have very serious consequences for our ever-growing world population.

But even for the home gardener, bees in the community can mean the difference between a meager harvest and a bumper crop. Fruits on properly pollinated plants tend to be fuller and bigger and more healthy, which means more to put on your plate.

REASON #3:
FORGING THE HUMAN CONNECTION TO FOOD

You've probably heard of chef Alice Waters, writer Michael Pollan, and maybe even writer and farmer Joel Salatin. Or perhaps you've seen films like *Food, Inc.* or *King Corn* that highlight the precarious nature of our current food system. These are some of the most well-known activists and projects pushing for more localized, sustainable food systems and fewer food miles or kilometres. (If you haven't heard of them, I suggest that any prospective beekeeper check them out.) As these speakers, writers, and films will tell you over and over again, many commercially available seeds have been genetically modified to withstand heavy insecticide use, farmworkers are constantly exposed

Few people get the opportunity to see honey in its unadulterated state, direct from the hive.

to dangerous chemicals that cause irreparable damage to their health, and corners are cut with regard to human and environmental health—all for the sake of the bottom line. To add insult to injury, the resulting crops are then trucked thousands of miles or kilometres or processed heavily and offered up to the public at prices so low that few people want to seek out better, fresher stuff. It's a dire situation, and humans are certainly not the only creatures suffering for it.

Over the past few years, the general public has become increasingly more aware of flaws like these in our current food system. But few of them know about the role of bees in commercial agriculture. Through the services that migratory beekeepers offer to industrial fruit and nut operations in the United States, honeybees account for an added $14 billion profit in the pollination of crops such as orchard fruits, berries, and nuts. In fact, our current food system could not sustain itself without the bees' hard work.

Migratory beekeeping is essentially the act of transporting large numbers of bee colonies from big farm to big farm, to fill in for a lack of pollinators due to the nature of industrialized, monoculture farming. On smaller, diversified farms or in the wild, bees find plenty of stuff to feed on: Weeds, grasses, and varied species of trees all with different blooming periods produce nectar and pollen, providing a balanced diet and a steady availability of food throughout the year. But on monoculture farms, farmers clear hundreds—if not thousands—of acres of farmland of native plant life and plant one singular crop in its place, such as corn or soybeans. The land is then heavily treated with pesticides and herbicides to guarantee the crop flourishes with little competition from weeds or pests. This means that any bees that have taken up residence in the area have a huge burst of one type of food for a short period and not much else. Not surprisingly, bees don't thrive under such conditions, and they will starve in the winter when no other food is available. To solve this dilemma, large flatbed trucks stacked high with honeybees are moved en masse from farm to farm according to bloom times. Once the blooms fade, the bees either are moved again to a place with other crops to pollinate or are fed a supplement of corn syrup or sugar as a substitute for nectar. Instead of natural pollen to help stimulate egg laying, they get soy flour. Not a very fair trade, if you ask me.

The number of hives being transported seasonally is staggering. I cannot count the number of times I've gotten a look of disbelief when I tell students that more than 60 percent of the beehives in America are transported thousands of miles across the country to pollinate just one California crop—almonds. The rest are moved around at various times of year to pollinate citrus fruits, apples, pears, and berries. But anything grown on a commercial scale—such as avocados, cotton, and soybeans— requires migratory beekeepers to bring in hives to efficiently pollinate the crops. Pollinated plants mean higher yields, which means more product to sell and more money in the bank.

As a beekeeper, there isn't a season that goes by now that I'm not aware of what is in season, not only for bee forage but for human forage as well.

Some beekeepers say that between the lack of adequate nutrition and the stress associated with moving hives long distances, more than 10 percent of the hives die in transit in each direction. These conditions have also been thought to be a contributing factor to Colony Collapse Disorder, or CCD, the mysterious malady that has claimed a frightening number of hives worldwide. Although there is still no confirmed cause for CCD, commercial beekeepers are often the ones footing the bill for the loss. They lose their bees at mind-boggling rates and have to scramble each season to replace them. To add to the pressure, they have pollination contracts they must honor for fear of going out of business. So they split their surviving hives or buy packages of bees to replace the bees that have died. And they meanwhile continue to truck their hives to far-flung destinations season after season, only to end up cleaning out dead bees again and again.

It's a difficult situation for these bee-keepers, but many of them are pushed to manipulate their bees in unnatural ways so that the job gets done. As a beekeeper, I feel empathy for these men and women. Beekeepers love their bees, and it is because of their hard work and sacrifice that our pantries are kept full. But it doesn't have to be this way: As people continue to fight against commercial agricultural models that make these migratory beekeeping practices necessary, it is important that small beekeepers maintain healthy, natural colonies around the country to help ensure some population of genetically diverse, robust bees so that if our food system really does go "pear shaped," at least we've still got some bees that know how to bee themselves.

Many beekeepers swear that handling their bees forces the keepers into a meditative state.

REASON #4:
CONJURING YOUR INNER HIPPIE

Urban apiaries offer city dwellers a much-needed opportunity to commune with nature in places where real nature is negligible. In New York City, for example, the parks are home to wild birds and beautiful trees, yes, but compared to rural towns, the cities seem limited mainly to trees growing in dirty sidewalk wells and pigeons nesting in the eaves of apartment buildings. Plus, because of efficient mass transit here and the high cost of living, many people lack the cars that make frequent travel to the countryside possible. Many of us are effectively trapped by the city that we so love, in other words.

So, during the spring and summer, our public parks fill up with nature-starved New Yorkers seeking to unwind and observe a little flora and fauna, maybe getting transported, if only for a short time, to someplace serene and unbridled. And many urban beekeepers will tell you that their bees help serve a similar function. Beekeeping is a way to feel connected to the natural world, and it tends to amplify what little nature your area may have. In fact, the bees will seek it out and bring it to your doorstep.

Working with honeybees can even be a form of meditation for some people, forcing them to remain present as they work, if only for fear of being subjected to stings. Slow movements and mindful positioning of your body and hands can improve your beekeeping, making it reminiscent of other forms of Zen practice. It can be tremendously rewarding to sit on the roof of your building at the end of a long day, beer in hand, and watch your bees come in from foraging as the sun sets. It's one of those moments that, once experienced, you wonder how you ever lived without.

What's more, the independence of the bee is especially appealing to people who prefer a low-maintenance, low-stress hobby. Bees, even as they live in manmade structures, maintain all of their wild instincts, and they can survive without human intervention under the right circumstances. A new beekeeper can effectively maintain an apiary in about fifteen minutes a week. And then at the end of each season, even the lazy beekeeper will reap the rewards of their commitment to overseeing the health and welfare of their colonies. As you're reading, you'll discover that those prizes come in ways you might not expect and are not limited to honey.

REASON #5:
STICKY SWEET PRODUCTS OF THE HIVE

Of course, honey is a pretty good reason for keeping bees. Nothing compares to the first time you harvest it from hives you've tended all season. It's an addictive moment and easily might be the main motivation for most hobbyists getting involved in beekeeping. Many beekeepers end up setting up shop to sell honey and other products of the hive to generate a surprising amount of extra income for their household. Here are all the goodies you'll be helping to create:

Honey

Technically put, honey is "shelf stabilized" nectar that the bees source from flowers and trees within a few miles of their colony. What does that mean? For starters, worker bees come back from foraging flights heavy with full stomachs. (You can often see them struggling to fly straight to enter the hive.) Upon entry, they'll pass the nectar onto the house bees, who will in turn deposit the sweet liquid into empty comb cells. Over the course of a few days, the bees sip and regurgitate that nectar back and forth into the cells, introducing enzymes and yeasts that help other bees digest it during the long, cold months of the year. (Honey and nectar serve as the carbohydrate source for bees, giving them abundant energy for all of their labor.) Those house bees will also fan their wings to increase air circulation, which helps evaporate off the excess water. When that happens, microbial activity ceases inside the thickened nectar, and the bees cap the cells with a thin layer of wax. Magical, right? Properly stored and ripened, honey in this condition will last indefinitely. It is the bees' equivalent to the human hobby of canning.

A healthy hive can potentially produce a tremendous amount of excess honey. Some beekeepers give estimates of 80 to 150 lb/36 to 68 kg. I find that to be a bit generous. In the first season, I generally only take a couple of combs from a new colony and leave the rest so that the bees will be healthier throughout

Comb honey is an exceptional treat, harvested right from the hive, protected in wax cells.

. But in the second season, it
ommon for a beekeeper to
nywhere between 50 to 100 lbs/
23 to 45 kg of honey from just one hive,
with plenty left over for the bees to overwinter with. That's quite a bit of sweet
stuff!

In many cities, truly local honey is both
sought after and difficult to find. Many
people like to consume raw, local honey
as a way to alleviate allergies; others
simply enjoy the flavor of their hometown.
At $5 to $30 per 1 pound/£3 to £20 per
450 grams, depending on locale and
availability, you can easily supplement
your income with honey. Small-scale
beekeepers can start their own honey
CSAs or sell their products to friends
and neighbors to support their hobby.

Beeswax

In addition to honey, the hive has
other equally useful substances that a
beekeeper can harvest for use at home
or for sale. One of those is beeswax,
or wax comb, which is made by young
worker bees inside the hive. Most, if not
all, of the structure inside of a beehive
colony is beeswax. Even at just two
weeks old, house bees have developed
mature wax glands, which are found on
the underside of their abdomens. With
help from a clear liquid they excrete
that solidifies shortly after exposure
to air, they can manipulate the wax
into the desired shape. Fresh wax is
very soft and often requires a little
time to cure, although bees will use
those fragile cells as soon as they are
completed.

Beeswax is usually harvested when you
extract honey. Once the honey has been
removed from combs, the remaining
wax is rinsed and rendered (see chapter 5
for the two techniques). That liquefied
wax is minimally processed—particulates
are filtered out—for an even, smooth
finished product.

Beeswax has a multitude of practical
uses: It is commonly used to make
sweet, fragrant candles (see Easy
Beeswax Candles, page 161), healing
salves and ointments, cleansing soaps
(see Honey Rhassoul Clay Facial Mask,
page 159), and emollient wood polishes.
It can be used in the kitchen as a sealant
and preservative. It can be used in
cheese making, mushroom cultivation,
and the bottling of syrups, tonics, or
liquors. Beeswax is also popular in arts
and crafts. You can make crayons for
children, make wax threads for beading
and sewing, or do etchings and lost-wax
casting . . . all with the wax left over from
the honey extraction process.

Pollen

Essential to plant reproduction, pollen
is a powdery substance used by angio-
sperms to fertilize their flowers, allowing
them to produce fruit and seed. This
is the plants' method of propagating
themselves not just so we can eat their
fruits but also so that their species can
continue to thrive in the world.

Honeybees collect pollen as a source
of food. It is the protein-rich ingredient
that they use to make what we call
"bee bread," a fermented pollen cake
combination of pollen, nectar, and

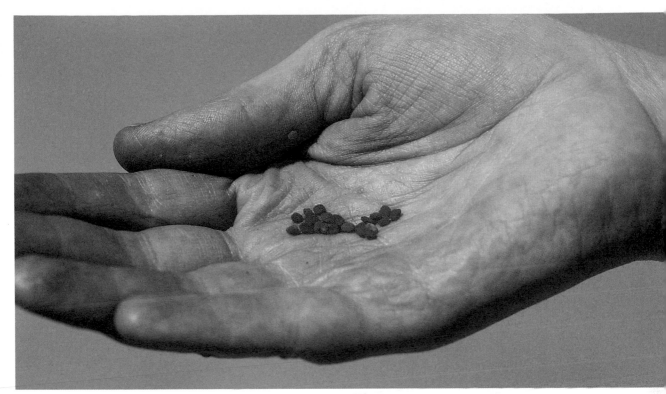

Tiny beads of pollen as they would appear on the hind legs of foraging worker bees.

glandular secretions eaten by nurse bees, which they covert into royal jelly. During foraging flights, the bees pack the granules of pollen into little beads using a flattened part of their hind leg called a "pollen press." The bees moisten the beads with a bit of nectar to bind them together for transport, put them into "pollen baskets" also found on the hind legs, and fly them back to the hive. Once home, the bees pack the pollen into wax cells and allow it to ferment, a process during which beneficial bacteria and yeasts grow. There, it will sit until it is needed.

But beekeepers have a brief opportunity to harvest some of the pollen for personal use. Using simple pollen traps available through beekeeping supply companies, the beekeeper can harvest a couple days' worth of incoming pollen without creating much of a dent in the bees' food stores.

Bee pollen is widely used as a nutritional supplement. While many of the claims about bee pollen have not been scientifically proven, studies have confirmed that pollen is packed with nutrients that the human body can absorb easily. Pollen composition will depend on the plant from which it was foraged, but typically pollen will be 25 to 35 percent protein and 55 to 65 percent carbohydrates, with the remaining percentage a combination of vitamins, minerals, and other compounds. As such, many people view bee pollen as a natural way to boost their nutrient intake. (Please

Bees use gummy, resinous propolis to seal the cracks between frames within the hive.

note that people with allergies should be cautious when ingesting bee pollen.)

While pollen may have health benefits, it is also has culinary applications. Pollen often has a sweet and tangy flavor that is reminiscent of the plant from which it was foraged. Chefs prize specific types of pollen such as that from fennel and dill, using them as delicate and distinctive flavoring elements. Try sprinkling some on fresh goat cheese with herbs or adding it to homemade pork sausages, breads, and pastas for beautiful golden color and unique flavor.

Propolis

One of my personal favorite things to harvest from the hive is propolis, or what beekeepers call "bee glue." Made from the foraged resinous sap from nearby trees, this substance is used to seal cracks and sanitize the inside of the hive. Since tree sap typically has antibacterial and antifungal properties that help keep the tree itself in good health, it's the perfect substance for bees to use to keep things clean within the hive.

Propolis is easy to identify. It's often a gummy brown- or red-colored substance that you'll see accumulated in the inner cracks of the hive. It's deposited in between frames by worker bees and can make it difficult to remove frames without significant disruption to the bees. Many beekeepers try to scrape off much of the propolis with a hive tool during inspections, which makes removing and replacing the frames much easier. You can harvest propolis this way by simply placing the chunks of tacky resin into a jar or baggie as you remove it from the sides of supers and frames. (Taking only what you gather during cleanup, like this, is my preferred method of harvesting it.) You can also use propolis traps, which you can buy from beekeeping supply companies and will find easy to use.

Once harvested, propolis has some pretty great uses. Some people will grind the hardened substance into a powder and allow it to dissolve in turpentine, using it to stain and seal wood. The coloration of the propolis stain can complement wood grain nicely and can vary depending on the time of year and the location from which it was harvested. And one of my favorite uses of propolis is in tincture form, steeping it in whiskey or moonshine. The end result is a strong medicine that works wonders for colds or the flu and is a great alternative to antibiotic ointments. Swab a little on a cut or scrape, and you'll be good as new in no time. You can even mix a teaspoon of the tincture with warm water and use it as a mouthwash; propolis has been known to help improve oral health and can be found in some toothpastes. Beware: Using raw propolis on your teeth results in an unattractive temporary staining of the enamel. Be sure to dilute it.

As with pollen, the potency of propolis is dependent on the botanical sources that the bees are relying on to make it. If you are curious about the origins of the pollen, propolis, or nectar in your hives, check out page 168 for lab testing services, through which you can purchase a detailed analysis of your bees' hard work. These reports are helpful not only for determining floral sources but also for checking for any unwanted contaminants. Bees are usually quite smart about where they harvest from but in desperate times, such as when the weather is extremely arid, they will opt for a manmade food source when natural sources have dried up. Sometimes they are opportunists and not connoisseurs.

Royal Jelly

Nurse bees—those responsible for caring for brood, or young bees—will begin eating the fermented pollen cake, or bee bread, and converting it into a milky liquid known as royal jelly when they are just three to twelve days old. Royal jelly is a glandular secretion that comes from the head of the bee and is deposited in the cells along with newly laid eggs. Once the eggs hatch, the bee larvae will feed on this substance. Depending on its status in the hive, a larva will receive more or less royal jelly: The larval queen, being at the top of the chain of command, will get the most of this nutrient- and enzyme-rich substance, which speeds up her development significantly. Other larvae will feed on it only the first two to three days of life.

Beekeepers who harvest royal jelly for a living have a method of procuring it that involves placing grafted eggs into a recently queenless colony. The young bees frantically begin feeding the new eggs royal jelly to create a new queen. After a day or two in the hive, the beekeeper removes the frame of grafted eggs, removes the small larva, and harvests the royal jelly. While it sounds kind of cruel, the number of larva destroyed during the processes is comparatively low when you consider the sheer number of bees being produced by a colony on a daily basis.

Almost every culture has accepted age-old medicinal applications for royal jelly. Whether being used in skin creams for reported antiaging properties (it's high in amino acids, which are proven to improve collagen production and skin elasticity), as a fertility booster, or as an aid in reducing cholesterol levels, people who use or consume royal jelly swear by its regenerative and healing benefits. Currently most of the evidence is anecdotal, but scientific studies are being done, and it's likely that much of what is believed about royal jelly is rooted in some truth. So while royal jelly should not be looked to as a miracle food, there is no doubt about the viability of it as a supplemental food source.

Bee Venom

Various cultures have used bee venom for healing purposes for ages. It's a powerful substance that is believed to have the ability to counter inflammatory disorders and autoimmune disease. As fascinating as the prospect of using bee venom for healing is, however, I cannot encourage people to go out and get stung by bees so that they are imbued with special bee powers or cured of whatever ills they suffer. Bee venom is

potent stuff; you should consult with a specialist before experimenting with it as an alternative type of healing.

In general, exposure to bee venom, or apitoxin, isn't dangerous for most people. That said, it is venom; there's no telling how a person may react to it, even if he or she has been stung before with no serious effect. When stung by a bee, most people suffer from some redness, pain, and swelling. A rare few (less than 1 percent of the population) will go into a systematic reaction, called anaphylaxis, to a bee sting in just a matter of moments, so messing around with bee venom as a novice could potentially be a fatal mistake. Leave this stuff to pros like the apitherapists, who treat painful disorders such as arthritis and fibromyalgia with bee venom using a bee stinger as a sort of acupuncture needle on pressure points. This technique is usually something they've learned by studying ancient medicine, and they often have many years of experience in holistic healing and acupuncture. Please don't experiment with bee venom on your own! Seek out an experienced apitherapist to help you.

At this point, you wouldn't be crazy to ask, "Why would anyone be interested in keeping bees at all if there's the risk of death?" In short, I will just say that while the risk is quite small, someone will always develop a moderate to severe reaction to bee venom. In situations where people have been stung many times in the past, they may end up developing a more intense immune response to one or two stings. If you think you are at risk for a more dramatic reaction—or you just want to be cautious—speak with your doctor or allergist about your plans to become a beekeeper and see if he or she has any suggestions. I have beekeeping friends who carry EpiPens and have even gone so far as to undergo controlled treatments so that they no longer suffer from systemic reactions to bee venom. This seems like a drastic measure to take, but these are folks who can't imagine life without their bees. People do crazy stuff for love!

Of course, prevention is always key. In chapter 4, I'll discuss in more detail what happens when you get stung and methods you can use to keep it from happening often. But as a beekeeper, it's usually a good thing to get stung by a bee every now and again. Your body can begin to develop a tolerance for the venom in very small doses, and it can even be good for you. But you don't want to overdo it. With patience and good technique, you won't be stung very often . . . just enough to keep you honest.

Now for the good news: Bee venom actually has some reported benefits for those on the business end of the stinger. Produced in two glands in a bee's abdomen beside their stinger, bee venom's anti-inflammatory properties (once a person's histamine response ends, that is) are frequently used to help treat rheumatism and several auto-immune disorders. Historically, cultures have used the bee stinger itself to administer the venom directly, but in today's market, many products containing harvested and purified bee venom can be used topically in a safe manner.

More research is needed to confirm the effectiveness of this ancient folk remedy, but in parts of Europe and Asia, bee venom has long been considered a true medicine. While the science of apitherapy has fallen out of fashion in more cosmopolitan locales, in some parts of the world beekeeping is still very much part of the cultural fiber, and bee venom is accepted as a treatment for many ailments. If you are interested in learning more about apitherapy, see the Resources section (page 168) for recommended reading and links.

CHAPTER 2:

Know Your Bees

To an untrained eye, an open hive of European honeybees, or *Apis mellifera*, seems chaotic and maybe even overwhelming. But with practice, you'll learn to cut through the buzz of thousands of little insect bodies to see what's really going on inside. In fact, most bee-keepers would never admit it to anyone, but beekeeping isn't exactly rocket science. I usually tell new beekeepers that the secret comes down to just one thing: observation. The essence of beekeeping is interpreting what you see, hear, and smell—or what beekeepers consider indicators of a colony's well-being. To do that well, you have to have a basic understanding of how colonies function. You want to be able to identify the parts of a bee's body; recognize what healthy "brood," or growing young bees, really looks like; and understand the difference between a drone bee, the queen bee, and worker bees and how each should behave during different seasons of the year. That's what this chapter aims to start teaching you.

THE NUTS AND BOLTS OF THE HIVE

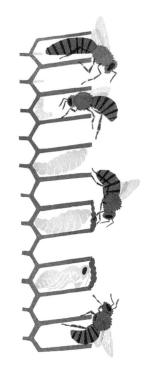

The inside of a honeybee hive is a fairly complicated place. Each hive holds one colony, and a colony consists of three distinctive types of bees: the workers, the drones, and a queen—plus the growing young bees, or brood, that will become more workers, drones, and, in rare cases, a queen. All of the colony members—and we're talking tens of thousands of them—are essential to its survival, and each of these three types of bees provides some vital function that contributes to the operation of the colony as a whole. Let's take some time to look at each type of bee individually, so that you can better understand how a colony works.

The Queen

Every colony has to have a queen. As the only reproductively mature female—a.k.a. the only bee that can create new bees—she is responsible for growing the size of the colony and, as a result, its capability to forage for food. That's why a healthy queen lays eggs as fast as she possibly can. If a colony does not have a strong matriarch at the helm, it can affect the bees' ability to thrive: Everything depends on the health and fertility of this one bee, and, as a result, she has to be one tough (bee) lady.

Despite her status, the queen begins life like all bees do—as a tiny egg laid at the bottom of a wax cell by the existing queen. Honeybees, like many insects, go through a process of development called complete metamorphosis. They begin their lives as an egg and then, after about three days, those soon-to-be honeybees break from their protective membrane and go into what's called a larval state. During that stage, which lasts four to five days, young adult worker bees called nurses (for obvious reasons) feed the larvae royal jelly and bee bread—a fermented mix of pollen, nectar, and bee's glandular secretions—and then cap the fast-growing insects in their cells with a mix of wax and propolis. This is now what's called the pupal stage, and it lasts sixteen to twenty-four days, depending on whether a bee is destined to become a drone (twenty-four days), a worker (twenty-one days), or a queen (sixteen days). At the end of the pupal stage, bees chew their way out of the cappings and emerge as full-grown honeybees.

But should the worker bees determine that a new queen is necessary (they are usually alerted to this need by weak pheromones sent out by an old or poorly mated queen), they'll give one egg copious amounts of royal jelly to speed up its development. Once that special bee hits the larval state, the worker bees begin building a large, peanut-shaped cell around the developing queen. When it's big enough, they cap the cell, and the queen completes the pupal stage.

On the sixteenth day, if all goes well, the hive has a new queen, and as soon as she emerges, she has an immediate mission: to eliminate the competition. She'll head out around the hive, making a high-pitched tooting sound, an activity known as "piping." She's trying to ferret out any rival queens so that she can destroy them, ensuring that she is the only matriarch in the hive. If that tooting call is returned by an existing queen or even a future queen still growing in a cell, the new queen will hunt her down

The queen bee, though difficult to spot, can usually be identified by a "rosette" of attentive worker bees attending to her every need.

for a duel to the death. Queens, just like worker bees, have stingers, which they rarely use except to dispatch their opponents. A queen's stinger is smooth and can be retracted easily and used again, unlike the barbed stinger of that worker bee that got lodged in my foot when I was a little girl.

If our virgin queen survives that battle, she'll spend a few days being fed by worker bees, finishing her maturation. After just two months, she is fertile and will take off on a week's worth of mating flights, called nuptial flights. She'll head out of the hive to a nearby "drone congregation area," the bee version of the local bar, a place where virile male bees are plentiful. Our queen will mate with several of the strongest specimens, storing as much of their genetic material in her spermatheca as she can gather in a few trips. Once she has mated with a half dozen or more (it's true!) over the course of several days, back to the hive she goes. And, after just a few days, she can start laying eggs.

A queen has the ability to selectively lay either drone eggs (which are unfertilized) or worker eggs (which are fertilized) and will deposit 1,200 or more of them on a daily basis, following chemical signals from her worker bees, which in turn respond to signals from her. As she lays, they will build cells depending on what type of bees the hive deems needed for the colony. This is revealing of the true nature of a honeybee queen: She is not really a leader but, rather, more like the heart of the hive—a very important organism inside a bigger one. She is resolutely and necessarily bound to

her task of laying eggs for the sake of the hive, and she only slows or stops that production when the cold of winter arrives, forcing the colony to slow down and conserve their resources.

A colony with a strong egg-laying queen (she's called "viable," in beekeeper speak) is referred to as being "queen-right." Fertile queens produce a series of very strong pheromones that, when sent out through the hive, suppress egg laying in worker bees, affect social behavior, and essentially assign chores, encouraging hive maintenance and queen cell building.

Beekeeping, in large part, is about ensuring that your colony is queenright. Though some beekeepers introduce new queens every season as a way to increase productivity, a well-bred queen can serve her colony well for two to seven years. (Hives having the same queen for seven years is rare, however.) And left to their own devices, your bees should eventually rear their own

Workers at the entrance of the hive, guarding against intruders.

queen when an existing one seems to be running out of steam. Even so, most of the time you'll spend watching your bees should be focused on assessing the quality of your queen and her egg-laying ability, as she's the heartbeat of your hive.

The Worker Bee

A young worker bee emerges from her cell.

The most abundant honeybee in any hive is the worker bee, or a "reproductively immature" female—a.k.a. a bee that can't lay fertilized eggs. They are everywhere, and, being smaller bees with almond-shaped eyes and fuzzy bodies, they are easy to identify. (Once you keep bees, you'll notice worker bees are larger when they return to the hive from foraging flights, their tiny abdomens swollen with their nectar haul!) Worker bees make up 85 percent of any colony, and for good reason: A queen is important, but the more worker bees you have to perform the duties the queen assigns to them, the more efficient your colony is, and the stronger and healthier it can become.

Worker bees are completely responsible for the oversight of all of the operations except for egg laying, and they perform

virtually all other duties both inside and outside of the hive, and each has different duties. They get their marching orders the moment they chew their way out of their cells, taking cues from the pheromones given off by the rest of the hive, including other workers, the queen, and even the still-growing brood.

First, they'll clean out the tiny wax compartment they emerged from, readying it for the deposit of a new egg or maybe food stores. From there, they might feed other developing brood or drones, fan their wings to manage hive airflow and temperature control, or attend to the queen. Eventually, they will make their way down to the hive entrance to help ferry incoming nectar and pollen to cells, stand guard at the entrance to help send the alert in case of intruders, or help clean out the bodies of their dead or dying brothers and sisters. (Worker bees live a few months at most; they are the only bees in the colony with barbed stingers and can use them only once, at their own peril.)

Eventually, a worker bee takes on the role of a forager. After a few orientation flights at about three weeks old, she will take off in any direction to look for water, nectar, pollen, or the tree resin used to make the propolis for sanitizing and sealing cracks in the hive. For about three weeks, she will forage tirelessly, until her wings weaken and her tired body gives out. During the busy spring and summer months, the entire life span of a worker bee is just six to eight weeks, while winter workers live much longer, as there is significantly less work to be done outside the hive.

If a colony has been without a laying queen for several weeks, worker bees will even start laying eggs. But because they never participated in a nuptial mating flight, worker bees lay only unfertilized eggs, also known as the drones, the last and lowest type of bee in the hive. When worker bees lay drones, that behavior, called laying workers, can put your colony in grave danger—and you'll soon understand

why when you read more about drones in the next section. A strong colony will police laying worker bees without your intervention, but in many cases a beekeeper should be sure to fix the problem. For instructions on how to deal with laying workers, see page 101.

The Drone

Ah, the lowly drone! Easily identified by its big rounded body, enormous eyes, and loud, clumsy flying, the drone is the bee that some beekeepers regard as a freeloader and a tramp. True, they aren't as handy as workers, but their presence is beneficial in a different way. Drones make up about 15 percent of the hive and are essentially the male of the species. As such, they're essential to bee reproduction.

For starters, drones are genetic copies of their worker mother, responsible for transferring her genetic material to other colonies during the spring mating process. That's when the number of drones in a colony will spike, as virgin queens in the area take their nuptial flights. The "guys" will gather in drone congregation areas to wait for a lovely lady to fly by. Once a queen is spotted, the drones make chase, and the fastest and strongest wins the chance to mate.

Gentle drones are easy to handle with no fear of being stung, as they lack the stinger that worker bees are infamous for.

Once the deed is done, the drones' reproductive organs remain lodged in the queen and, like the stinger of a worker bee, are pulled out of the bee's abdomen when he attempts to fly away. It's a sad end for those special few that ever do get to mate, but, hey, their gift of life (in two ways) is much needed.

Drones take the longest amount of time to develop—twenty-four days—and live for only about two months. In the meantime, they are waited on hand and foot by the workers. Drones can't forage for food, they don't assist with house-keeping duties, and they don't help with brood rearing in the hive. Most drones can't even feed themselves, thus the

belief that drones are merely drains on the strength and resources of a colony. Some beekeepers believe it's best to try and minimize the number of drones to increase the workforce, resulting in more foragers and more honey.

But I believe that in the colony, there is a symphony of pheromones at play. Each bee plays a part in the chemical song of a hive, including the drones, which give off their own pheromones. When the pheromones of the queen, the brood, the workers, and the drones are in balance—and food is coming in—the colony is at its best and is able to manage itself with little manipulation beyond the occasional inspection. In

A nice-looking frame of capped brood and honey, with bees "festooning" in an attempt to build more comb.

short, drones are good for colony morale and good for keeping the area stocked with genetics for breeding. I, for one, love drones. They are cute little guys, and, better yet, they don't sting!

All of this, however, doesn't mean drones should get a free ride in your hive forever. And they usually don't. Once the temperature drops and food sources dwindle, workers begin withholding food from them. They grow so weak that even the significantly smaller worker bees can forcibly remove them from the hive, leaving drones in the cold to freeze or starve to death. It seems cruel, but winter is when bees need to carefully conserve food. Drones are big eaters and put too much pressure on the colony's finite food supply, so out they go. And in the spring, the queen will just begin producing more drones to populate drone congregation areas with more males for future queens, continuing the cycle of the colony.

The Brood

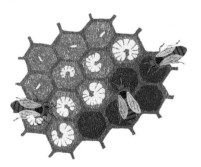

The name might sound like something out of an '80s horror movie, but the brood is a lot less sinister than it sounds. Simply put, *brood* means all baby bees—eggs, larvae, and pupae all fall under the collective category of brood. Brood begins as an egg the size of a pinhead laid by a queen in the bottom of a wax cell. For about three days, the egg is fed royal jelly until the outer membrane hatches and a tiny larva emerges, eating from the puddle of royal jelly that it lies in. Once it hits the larval stage, it continues to eat, growing to fill the entire cell in seven to ten days. The cell is then capped with a mixture of wax and propolis, and the larva develops through its pupal stage and emerges as an adult.

It's easy for a novice to mistake capped brood comb for the considerably more edible honeycomb. You can identify brood comb by its dry, opaque capping, which is usually reddish brown or even yellow. Honey cappings, on the other hand, are usually white or even transparent. You can more easily recognize uncapped brood because you can see the white, pearly larvae curled up at the base of each cell. Brood comb is also almost exclusively in the centermost part of each hive frame, with a little honey stored around the upper corners for easy access.

Spotting eggs can take a little more skill, but as you continue performing inspections, your eyes become trained to seek out the tiny flecks in each cell, which are about a quarter of the size of a grain of rice. The eggs will be standing up on end, making them even more challenging to spot, especially if they have been laid in virgin wax, as they are pretty well camouflaged against the clean white of unused new cells. As the

Note the difference between drone brood (top) and worker brood (bottom). Drone brood is larger, with rounded cappings, while worker cells are flat.

comb gets more heavily used for brood, it will darken, making eggs and larvae much easier to see.

A beekeeper should be able to identify the difference between worker and drone brood. Worker cells are much smaller, and typically very uniform; their cappings are flat and smooth. Drone cells, on the other hand, are slightly larger—about 25 percent bigger—than the worker cells and are easily identified by the more rounded, bubbly cappings that are built over the larva to help them continue their development. The health of your brood is also a great indicator of the health of your colony. Eggs and larvae should be glistening white, and worker cappings should be smooth and opaque. Brood with sunken cells and discoloration in larva and pupa can be indicative of a pest or a disease infestation. But if the brood nest appears healthy, your colony is likely healthy, too.

It's also important to pay attention to your brood to make sure your colony is queenright. Beekeepers often look for what's called a solid "brood pattern." If you see one egg in each cell perfectly centered, you know you are probably queened right. An ideal queen specimen will take full advantage of all available cells, filling up nearly an entire comb with brood. And if there are very few empty cells in between occupied cells and nearly the whole frame is filled, that would be considered a tight brood pattern. It's an indication that not only is the queen performing well but also the young worker bees are showing good hygienic behavior. That's because a queen will skip cells that seem insufficiently tidy, affecting her laying pattern.

If you encounter spotty brood, don't jump to conclusions too quickly. Spotty, loose brood patterns could also be indicative of a young queen learning the ropes, changes in weather, or a queen with weak genetics. If you can't find any new eggs at all and are finding only capped brood and older larvae, you may have to help the bees by re-queening. You can read about how to do that on page 97. But if you are seeing individual eggs and larvae in worker cells, there's still a good chance that the bees will be okay. In most cases, I've found the bees themselves will sort it out.

THE ANATOMY OF A BEE

Few creatures are as well suited to their roles as *Apis mellifera*, the European honeybee. From the moment these insects emerge from their cells, their anatomies are put right to use, evolving as they grow and take on new duties. Our current food system depends heavily on bees to perform the act they are practically designed to do, but now they're doing it on a large, commercialized scale. Bees work tirelessly at their assigned task literally until they have seen it done or they die, a trait that is both valued and aspired to by humans. Unlike us humans, however, bees are equipped with a multitude of tools to make their laboring efficient and seemingly effortless—in fact their anatomies make them true masters of their domain. Let's take a look at those worker bees part by part, outside to inside, so that you can better understand what makes them so darn good at, well, being bees.

The Exoskeleton

Most mammals, humans included, have a layer of flesh and hair over bones that conceals our fragile innards. But adult insects like bees have to protect their delicate vital organs with "bones" on the outside. This is commonly referred to as the *exoskeleton*. A honeybee's armored body is made up of plates of chitin (pronounced "kite-en"), which is the same matter that makes up our hair and our fingernails.

With *Apis mellifera*, many short, fine "hairs" also cover the entire exoskeleton. These hairs are actually an extension of the chitin that makes up their exoskeletons. They help honeybees collect pollen from flowers: An electrostatic charge builds up on hairs as the bees fly. When a bee lands on a flower, the pollen is magnetically pulled onto her body, allowing her to transfer pollen to flowers inadvertently as she roams. In this way, she is essentially responsible

for fertilizing female flowers, which, ultimately, produces the fruit and seed we eat and sow. Without those little hair-like protuberances, the honeybee would not be nearly as efficient at collecting the high-protein dust needed for pollination.

The Head

The head of a honeybee is like a Swiss Army knife—it's a marvel of nature. In this approximately ¾-centimeter-size noggin are several tools that the bee uses to perform many of the tasks it must do to keep the colony fully functioning.

First, on each side of a bee's head, there are two compound eyes that are filled with hundreds of smaller eyes called *ommatidia*. Each ommatidium contributes a little pixel of a larger picture of what the bee sees. The bee's compound eyes are used for exploration, observing, and deciphering its surroundings, much the same as other creatures' eyes. The bee also has three smaller simple eyes called *ocelli*. These are located on the top of the bee's head and are used for navigation and orientation. These eyes register the location of the sun, and with them and the help of landmarks "seen" by the compound eyes, bees are able to travel back and forth to a food source.

Another highly specialized tool on the honeybee's head is the antennae. Antennae are two short appendages on the bee's face, just above its compound eyes. These act as a highly sensitive scent receptor. Bees use their antennae to receive pheromones from their hive-mates; detect floral, resin, or water sources; and pick up on foreign scents from intruders such as animals, insects, or humans. Bees also use their antennae the same way humans use their hands: to touch. Below the antennae and eyes are the mandibles, which act like jaws. They're used for manipulating and molding wax, feeding brood, and moving objects in and out of the hive. Behind the mandibles is the proboscis, a retractable, tongue-like appendage that works like a straw. Bees insert it into a flower to retrieve and deposit nectar and also use it to feed young bees, cure honey, and drink water.

And finally, underneath those ocelli, is the brain of the bee. Made up of a series of nerve cells, the bee's brain takes in all this sensory information and conveys it to other parts of the animal's body. This brain might be far more simplistic than any mammal's, but it is able to process what's relevant to bees' needs with impressive speed, allowing them to perform with grace and determination.

The Thorax

Only slightly bigger than a bee's head and located between the head and abdomen, the thorax makes an impressive amount of functions possible. This is where bees salivate, and this is also where they breathe, thanks in part to the spiracles found both here and on the abdomen. The thorax is also the anchor for legs and wings, the latter being the most important feature. These transparent, delicate accessories are

A worker bee drinking up nectar with her proboscis.

responsible for transporting each bee over five hundred miles during its brief lifetime.

Honeybees have four wings total: two anterior (or fore) wings and two posterior (or hind) wings. A series of hooks, or hamuli, join the two wings together on either side and keep all four wings moving in synchrony. When in flight, about eight muscles in the thorax control it all. Instead of simply moving the wings up and down, the muscles help wings pivot and shift, to account for weight load and wind. A bee's wings move forward on a downbeat and backward on an upbeat, giving the insect even more precision in flight. With this specialized ability, and a wing speed of two hundred beats per second, honeybees can perform magical maneuvers such as hovering in place or changing direction in an instant.

Sharing the thorax with the all-impressive wings are six legs. Though they are most frequently used for walking, each pair performs specific functions, just like the hands and feet of a human. The first two sets of legs are used primarily for grooming the antennae and head, for passing pollen and propolis to the hind legs, or for carrying larger objects from the hive like dead bees or pieces of debris. The last set of legs, those closest to the abdomen, help to pass sheets of wax to the front legs and mandibles for use in building comb. These hind legs also possess a pollen press, with which the bee forms pollen into little compacted beads. The beads are stored in the pollen baskets, hairy segments on the hind legs that are used to transport the balled-up pollen back to the hive.

The Abdomen

The posterior of a bee is the biggest part of its body—and for good reason. The majority of the bee's vital organs are contained within these segmented exoskeletal walls. These segments, called *ventral plates*, allow for both flexibility and durability. And beneath them are the bee's digestive, cardiovascular, and reproductive functions, as well as the home of wax and pheromone production.

In fact, a bee's body is filled with hemolymph, a fluid with a similar function to blood. While hemolymph in honeybees is not responsible for the movement of oxygen like our blood is, it does help to distribute nutrients to the organs. That happens via the honeybees' simple circulatory system: a small dorsal heart located in the top lower portion of the abdomen and an aorta that helps transfer hemolymph to and from the brain and other parts of the body.

The bees' digestive system is also fairly simple. It begins in the head with that straw-like proboscis, which leads down the esophagus through the thorax to the abdomen, where just over the threshold sits the bee's crop, or "nectar stomach." That's where forager bees store water and nectar, filling it so that it expands to full capacity before they return to the hive to regurgitate the contents into a cell inside. Any nectar that isn't regurgitated into a cell is eventually passed to what's known as a midgut, where it will be digested further and eventually excreted. Pollen is digested in the same manner, though it isn't stored in the crop en route to the hive.

In the lowermost part of the abdomen are the stinger and the venom sac. When threatened, bees use a series of muscles to extrude a barbed lancet, normally sheathed within its tail end. When the stinger penetrates an invader's flesh, venom is injected into it, and, in many cases, the stinger stays with it. As the bee flees, the stinger and venom sac are ripped from its tiny body, after which the bee will fall to the ground and die. Meanwhile that remaining venom sac still pulsates, pushing more venom into the offender, a gruesome touch that increases the effectiveness of the sting. As a result, any critter with a decent memory will quickly learn to associate a beehive with hours of pain.

While some bees meet this sad end, it is usually not in a colony's best interest to mount a full-on attack on an intruder. The loss of the workforce can weaken the colony and lead to lower productivity, a less-organized and fastidious hive, and an inability to keep up defenses against pests and future intrusions. That means bees use their stingers only when they feel their colony is in danger. Most worker bees will live out their lives without ever using their stingers.

On warm, sunny days, workers jump at the chance to hit the skies in search of food.

A HIVE FOR ALL SEASONS

Like everything in the animal kingdom, the behavior of honeybees is dictated by food and how much of it is available, which is, of course, dictated by the seasons. You can usually connect anything going on within the walls of your hives to what's happening outside. Here's a description of what happens in a bee colony, season by season.

Early Spring

In late March—or as early as late February, in some regions—the bloom season begins. Trees are among the first plants to blossom, and they are a saving grace for honeybees after long, cold winters. As temperatures begin to hit 60°F/16°C, the bees break their

Workers ripen foraged nectar in cells to get it to a desired thickness before capping it with a thin layer of wax.

cluster and start foraging for pollen and nectar to begin the new year's brood-rearing process. The pollen and nectar that the trees provide is abundant, and it is constant throughout the spring. The queen fattens up and begins laying eggs at a tremendous clip because the colony needs to build up quickly to take full advantage of the abundance of spring food. More foragers in the field (literally) means more resources coming in and more that can be put away for winter. And that means the bees have a better chance of survival for yet another year.

The colony also produces a higher volume of drone brood in order to saturate their area with male genetics when swarm season begins. During swarm season, the rapidly growing hives will break ranks to look for new places to create colonies. As the days get warmer, drones emerge from their cells and hang around the hive eating and building up strength. After two weeks of this adulthood, they reach reproductive maturity and head out on flights to scope out the local dating scene.

As the peak forage time arrives, space gets tighter inside the brood nest. Pollen, that protein-rich food necessary for creating the royal jelly fed to developing brood, is coming in quickly, and the rate of baby-bee production spikes. As a result, the first few weeks of the early-spring nectar flow are when urban beekeepers with overwintered colonies might need to prevent crowding in their brood nests to keep them from swarming. We cover this below and in chapter 4.

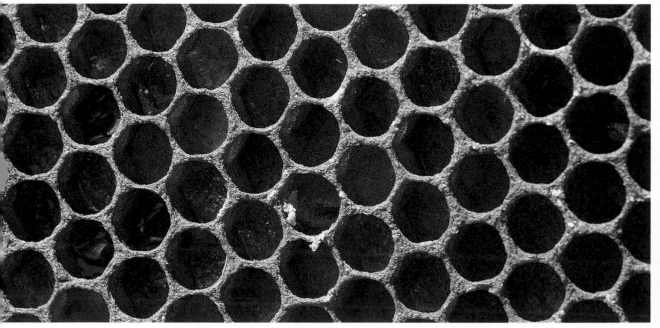

Bee bread, a mix of pollen, nectar, and glandular secretions, is packed into cells to store until it's ready to be eaten by young bees.

A full frame of honey, such as this one, is ready to harvest once it is nearly completely capped.

Early spring is also the time to start new hives, and most commercial beekeepers sell bees from April to June. To find out how to buy them and start a new colony, see page 75.

Main Season

By late April, the hive is in full swing. As soon as an adult bee emerges from a cell, the queen lays another egg inside. The workforce is expanding fast, and the hive can continue to get crowded. At this point, a beekeeper needs to give the bees more space to continue building comb and expand their community. Without more space for brood production, the bees will prepare to swarm, which means a group of bees will head out en masse to create a new division. This behavior typically occurs only with overwintered colonies that come into the season strong in number, though in cases of neglect in new colonies, swarming can also take place. Most practiced beekeepers have some tricks up their sleeves to prevent this from happening, because divisions result in fewer bees in the hive and, as a result, a lag in productivity. While swarms are a bit of a spectacle and can startle people who don't understand what is happening—hundreds of bees gathering on a limb or signpost—swarming bees are actually at their most docile. And beekeepers can respond with some gentle and effective methods of managing swarming, without completely suppressing the bees' natural urge to multiply, which we will explore fully in chapter 4.

Once your bees have decided to stay put, they will continue to build comb in the upper parts of the hive to use for storing nectar. Workers will build comb—beekeepers call it drawing out the comb—for as long as there is more food and brood than the current comb can house. That's why beekeepers must make periodic inspections to make sure the bees have enough space for more brood and honey.

By high summer, the bees have started to put away a bit of excess honey, the amount of which is directly proportionate to the strength of your colony. If stores are strong, it's a good time to harvest some honey and replace those full combs with empty ones ready to be built out with comb and filled in with more nectar.

Late Season

At the end of the season, during the summer's hottest months, flowers start to dry up and die, resulting in what beekeepers know as a "nectar dearth." It usually happens for those few weeks in between seasons when trees stop flowering but the grasses and weeds haven't begun to bloom. With food sources waning, bees can try and rob weaker hives or forage off less-desirable non-floral sources like hummingbird feeders or half-empty soda cans. You can prevent your own hives from getting robbed by shrinking the entrance to the hives of weaker colonies. This will make it easy for the bees to defend themselves. It is also important to remove any spilled honey and clean up your hive immediately after you harvest. Any source of syrupy sweet food, in fact, may end up attracting bees from other colonies to yours.

During an especially long nectar dearth, you might want to feed your bees—in the form of sugar syrup or honey—to help suppress the urge to rob. Before adding a feeder (see "Hiving a Package," page 79), however, inspect your hive and be sure to mark the frames that are capped or nearly full. You will want to avoid harvesting honey that is contaminated with table sugar.

By late fall, it's time to prepare the colonies for winter. The bees will naturally start moving nectar down into the lower part of the hive, or the brood nest. And you will want to reduce the size of the hive by removing any excess frames of honey, or supers. One full box filled with frames of honey on top of a brood nest is usually ample food stores for bees in winter, but it's a good idea to make sure you've got at least four full frames of honey per level, or 60 to 80 lb/27 to 36 kg.

Late Winter

While solitary bees hibernate in the coldest time of year, honeybees do not. Summer-born bees will cluster inside of a hive, packed loosely into a ball around the queen, gently vibrating their wing muscles to generate friction and warmth. Depending on how cold the hive gets, the cluster will expand and contract.

During feeding, the cluster will also migrate slowly over the stored food. Bees located at the center, closest to the frame, will take up food and circulate outward, sharing some with their sisters as they move to the outside of the cluster. This creates room in the warm center for the colder bees on the outside to move inward and warm up.

While winter worker bees can live between four to six months, their main purpose is essentially to be a warm "body" to get the colony through the coldest months of the year. They differ from spring bees in that their hemolymph holds more proteins and fats, which help them to survive winter's chill.

At this point, almost no brood are being reared in the colony, as the low temperatures would kill the fragile eggs and pupae. Brood rearing usually requires an outside temperature above 50°F/10°C and an incoming source of pollen, so at this time of year, the bees are just sitting tight and waiting for the first blooms of spring.

On milder days when temperatures do hit 50°F/10°C, the bees will break cluster long enough to go on cleansing flights and forage for water. Honeybees are fastidious creatures and will not defecate inside of their hive unless they are ill. Because of their confinement during the cold months, this is often when most hives collapse from disease or starvation.

Most beekeepers check on their hives in February to see if honey stores are still plentiful and to confirm that the hive is still alive. If the hive is low on stores, the beekeeper will offer the bees emergency feed in the form of stored honey taken from hives in the spring, or sugary fondant icing, which the bees can easily consume in low temperatures. If the bees have not survived, the beekeeper will clean up "dead outs" and prepare the woodenware—the components that make up the beehive—in anticipation of new bees in the spring.

CHAPTER 3

Hives

Now that you have a fundamental knowledge of the honeybee and how the colony and beehive work, maybe you are making plans to buy a hive or two of your own. I often say that being a successful urban apiarist boils down to a combination of hive placement, good neighborly relations, and a little bit of commitment. But of these three factors, I believe the most important element by far is the placement of the hive. It's the foremost factor for most people thinking about keeping bees, especially if you don't have any yard space or if you rent your apartment or home. And then, even after you've settled on the "where," you've got a whole other set of variables to consider, such as what sort of bee setup you'd like to buy and what sort of hive you'd like to use. But don't worry, we'll cover all those matters here.

LOCATION, LOCATION, LOCATION

Most people find a way to have bees if they want them: the backyard of a friend with an understanding land-lord, a restaurateur with roof access, a community garden. But you can't put a beehive down in any old corner and expect it to thrive. It helps to run down a checklist of criteria for a proper hive location. Where your hive is situated can make the difference between a hearty, productive colony and a weak one that may not make it through its first winter. It can also mean the difference between an enjoyable and inspiring learning experience and a season of headaches and hassle; and moving a beehive once it and the bees have already been situated is not an easy task. So before you place your hive someplace, consider the following criteria.

Sunlight

I firmly believe that the single most critical aspect to keeping bees healthy is finding a place with adequate sunlight. Ideally, your bees will get direct sunlight early in the morning and then as much sun as they can after that. Many beekeepers also recommend dappled light in the late afternoon, and this is valid, but in my experience, I've found that the more sun the hive gets, the better. Shaded hives can have a bad attitude.

Bees work by the sun, and if they don't feel it shining on the face of their home, they are slower to get to work in the morning. Less sun means less-productive bees, which means less food for the colony, which means it grows more slowly. Fewer bees also tend to keep a less tidy home, which can lead to disease and eventual collapse.

In fact, sunlight is one of a few ways that urban rooftop colonies have an advantage over backyard or rural hives. At sunrise, the sun's first rays of light shine on the well-placed urban apiary, warming it and telling worker bees it's time to get down to business. (If you plan to put your hives at ground level, make sure the spot you've chosen will get the maximum amount of sunlight, and as early as possible.) Consider this the bees' version of getting up on the right side of the bed each day. From spring to summer, the workday of a city honeybee can sometimes last until eight-thirty at night, while the sun is still glowing softly on the tops of tall buildings. This wide window of oppor-tunity each day can help contribute to a healthier colony—and a healthier colony can produce more honey for themselves and, eventually, for the beekeeper. This is not to say that if your hive is in a sunny spot in a big city, you are guaranteed to harvest a bumper crop, but abundant sunshine will give your hive a good start.

You'll also want to face the hive south to southeast so that the sunlight shines right on the entrance, warming the portal that the bees will pass in and out of on their foraging trips. Will your bees fail if they are facing the wrong direc-tion? Of course not, but it's the way I was taught, and it hasn't seemed to hurt the bees, so I encourage that other folks do

Situating hives near a chimney or other permanent structure can shield them from wind and sun.

the same. But if your sunniest hive location doesn't allow for facing the bees southeast, face them in the direction that works best for you.

Last but not least, keep in mind that you will need a clean water source designated for the bees available at all times, as they will need it to help to cool the hive during particularly hot weather, so your location should allow not just for sun but also for water.

Safety and Stability

Sun is important, but to spare yourself a lot of trouble in the future, it's imperative that you pick a location that is safe for you, your neighbors, and the bees themselves. First, if you plan on keeping your bees on a rooftop, you must determine that the roof can support the weight of the hives in mid- to late summer when they can weigh 200 lb/90 kg or more each. On the ground, this isn't a big issue. You'll just want to make sure the ground is solid with good drainage. But rooftops require a bit more consideration, and you also have to account for your own weight when you are inspecting the hives, plus that of any other curious guests you might have accompany you. If the roof can't support that amount of heft and then some, look for another location.

Check with a roofing contractor or an architect friend you can trust if you want to be certain the location is structurally sound. Distributing the weight onto pallets can also help relieve pointed pressure, which can damage rooftops by puncturing weatherproofing or causing sinking in the structure itself. If your hive tips over and bees are spilled out everywhere, they may end up dying from exposure or just scaring the tar out of your neighbors.

You also have to consider the level of exposure to the elements. Have you ever had gusts of wind or flooding? Does the temperature get so hot that you yourself cannot sit on your roof in August for a significant amount of time comfortably? If that's the case, try to position the hives near a parapet or other windbreak or even in the shadow of a chimney that might also give them some respite from the blazing sun. You could also add a self-watering garden planter. These will not only serve as a wind barrier and source of shade, but they will also double as a water source. Bees are tough, but rooftops aren't usually a first choice for feral bees, so try to make things easier on them by minimizing environmental stresses.

Once you've got a good idea of where your beehives are going to go, you'll need to do a final bit of fussing. You'll want to situate your hive in such a way so that the entrance of the apiary faces away from the flow of traffic—human, animal, or machine. (Motorized machinery used in close proximity to a hive can cause vibrations and put the bees on edge.) During the spring and summer, worker bees by the thousands tear in and out of the hive looking for food and water. They return, heavy with resources, only to head out again after dropping their loads inside the apiary. If you stand back and observe a colony during the late spring and early summer, when it is at its busiest, you can see a clear area of activity around a hive. It's usually suggested that a colony should have at least 10 to 15 feet/3 to 45 metres of clearance at the entrance to allow for an unencumbered flight path, and some beekeepers suggest putting up a barrier in front of the entrance—in the form of a shrub or a wall—just to force the bees to fly above the path of human traffic.

And then think carefully about your community: Will your bees' normal flight paths interfere with the day-to-day life of you or your neighbors? Will animals or children play within 20 feet/6 metres of your hive? If you answered "Yes" to either of these questions, explore other options for a location. Beekeeping is a safe hobby for older kids to take part in with their parents, but if the children aren't yours, it's always better to be safe than sorry.

Remember it's not so much that bees will automatically cause a problem: Bees are remarkably docile during their foraging trips. When out on a field run, workers are so focused on the task at hand that they scarcely notice when humans are outside of their home. But bees can easily get tangled in hair or fly up a pant leg or into a shirt. Once trapped, a bee's stinger comes out, and it's game over. Accidents happen, so make sure that area around the front of the hive stays clear.

On gusty locations like rooftops, be sure to secure hives with straps or a cinder block atop the outer cover.

You should also consider that your location may be subject to gusts of wind that could rattle the beehive structure. If the hive faces that possibility, you may choose to ratchet strap the hive to its stand or set a heavy rock or cinder block atop the outer cover. You may even want to place the hive near a windbreak such as a chimney tower or a low wall. Either option will make it a bit easier for nectar-laden foragers to land at the entrance and keep gusts from blasting the lid off of your hive. True, I have never heard of an urban hive being bowled over the side of a building in a gale. And bees do a pretty fine job of stabilizing their hives with the propolis they make themselves. But it would be wise for a beekeeper to take an added precaution, just in case.

There's no such thing as being too prepared as far as urban beekeeping is concerned.

If you are using a pallet or a hive stand, you'll also want to make sure that it is level before putting your hive down on top of it. Bees use gravity as a guide for comb building in a behavior beekeepers call "festooning." That's when bees link together in a chain, hanging from the underside of the top bars of a hive, almost like the way folders hang in a filing cabinet. That way the bees know how to build a comb straight up and down—by letting gravity show them the way. You will see this most often with hives that are allowed to build natural comb without the aid of a foundation.

If your hive is level, bees will festoon and build the comb to fit perfectly within the frames, even connecting the base of the comb to the bottom bar. If your hive is tilted, the combs get fused to adjacent frames. The result is a mess, and your hive is hard to inspect without cutting the comb apart. You can avoid this by simply putting a level on top of the hive stand and balancing the hive using wooden shims or clay tiles, a project even easier than hanging a bookshelf. Some people also like to raise the hive about ¼ inch/6 millimetres on the back side so that it sits on a slight incline; that way, any moisture that accumulates in the hive rolls forward and out of the entrance. This keeps the bees dry during the colder months of the season when collecting condensation can be a problem.

Accessibility

Unlike a garden, which you'll want to check on every day, you want to leave beehives alone for the most part. There is a fine line between "fussing" and "neglecting" in regard to honeybee management. That said, you are going to need to check your hives from time to time to make sure that everything is going smoothly. I can tell you from personal experience that if your hives are a pain in the neck to get to—far away, in somebody else's locked yard, or in a precarious place that actually makes you nervous to get to—you are going to be less inclined to perform routine inspections. You might not mind one-armed ladder climbing and squeezing through windows, but just use common sense when navigating more-challenging apiary sites. It's kind of like going to the gym: If it's easy to get to, you're more likely to go.

Bees left to their own devices can fill their hives up with combs in a hurry. In a human-made hive, that means bees can run out of space to expand their nests or store food. And when that happens, they often begin planning to swarm. In the country, swarms are often just a hassle, because no beekeeper wants to lose any bees, but in the city they can spook your neighbors and create real drama. (In Brooklyn, where I live, one New York City police officer has been nicknamed Tony Bees because of his expertise in dealing with swarms.) Routine inspections will give you the chance to add space, when necessary, giving your bees all the room they need.

Another reason to perform inspections with some frequency is to make sure your colony has a viable queen. This is vital during your very first season when your bees are still getting established. If a hive is queenless, that means no new worker bees are being created. Sooner or later, a queenless colony will collapse if a new queen is not introduced. The longer a colony goes without one, the more slowly they'll grow to a sufficient size, and the less chance they'll have of making it through their first winter. In fact, queen-lessness is one of the biggest threats to bee survival, and it is one more reason to make sure you can get to your hives every couple of weeks.

Your bees might also get sick or infested with pests, and if you aren't inspecting them regularly, these situations can easily get out of hand. What's more, without regular inspections, you may not even have enough experience watching normal bee behavior to know that they are sick in the first place. When you take the time to observe your bees on a semiregular basis, you'll start to get a feel for what constitutes a normal behavior and what is abnormal.

That being said, there is such a thing as inspecting your bees *too* often. Each time you open the hive, you undo some of the work the bees have done to propolize gaps, build wax, and regulate the hive's temperature. And their symphony of pheromones is disrupted as air wafts through the hive. Limiting the intrusions to once every couple of weeks is a good idea, in my opinion. One inspection weekly is as often as I'd recommend during your first season.

Of course you might also want your hives to be someplace where you can share your experiences with friends and neighbors. You obviously want the hives where the risk of someone being hurt in a non-bee-related incident would be low—a rooftop accessible by stairs with a retaining wall or a ground-level community garden—rather than a rooftop that requires you to shimmy up a ladder. If you are going to invite visitors to come see the hives, consider their safety. You might also want to consider protecting yourself legally by requiring visitors to your apiary to sign a waiver before participating in an inspection.

Access to Water

A lesser but critical element to consider for placement of your bees is a water

In many cities, roof access comes in the form of a hatch in the ceiling. Make sure it's safe and that your gear will fit through the opening!

source. As mentioned earlier, honeybees use water to cool the hive during the warm summer months and will forage it in the place where it is most abundant. Oftentimes a drippy spigot, a small fish pond, or a neighbor's swimming pool will stand in as a watering hole if you don't set up a designated water source for them before the start of the season. Water buckets weighted down with rocks (and complete with sticks functioning as little life rafts for drowning bees) are one easy way to provide water, but five-gallon/twenty-litre galvanized chicken waterers available through online hatcheries and farm supply stores are a personal favorite.

Hives that receive more sun tend to be more productive, and have fewer issues with excessive moisture build-up, which can lead to disease.

Since they are enclosed, evaporation won't happen as rapidly, and there are far fewer instances of drowned bees. Choose something large if you can and place it 15 to 20 feet/4.5 to 6 metres from the hive so that the bees are more likely to come into contact with it as they exit on a foraging trip. Make sure you fill the water source often. Once it dries up, the bees will begin looking for a new one, and it might end up being someplace where they can congregate in large numbers and become a nuisance. To help entice bees to the source, you may choose to put a few drops of aromatic lemongrass oil in the water until the bees get accustomed to visiting their assigned source.

Rooftop hives without an accessible water source can have a harder time keeping the temperature regulated, which can result in collapsed combs or the entire colony absconding. When bees can't perform adequately in one location, they sometimes will choose to leave and start over again someplace more desirable. It doesn't happen often, but it can happen. Keep your bees from even the possibility; make sure they have access to an abundance of clean water.

Good Neighbors

I have always had a policy of honesty with my neighbors as far as my beekeeping endeavor is concerned.

I've never felt as though keeping bees in the city was wrong, so I didn't feel compelled to hide my interest in it. Before I set up my hives, I had a conversation with my closest neighbors about it, and after explaining that I had taken classes and would have help from other beekeepers and that it was safe, they all agreed that it would be fine to put some bees on the roof. Perhaps they didn't know what they were getting themselves into, but even after five years of keeping bees at my current place of residence, there have been zero complaints. I am fortunate in that I live in a neighborhood occupied by a large number of Polish immigrants. In Poland, beekeeping is still a common practice, and, in fact, hobbyist beekeepers are everywhere in Europe. I assume that the sight of a beehive didn't seem that weird to them. A couple of my neighbors even proudly shared that they have beekeeping relatives and recounted fond memories of them with me. I am extremely lucky to have been met with such open-mindedness. My neighbors told me that if at any point the three hives on our roof became a problem, they'd let me know. That's a totally reasonable agreement, I'd say.

It bears mentioning again that I had a friendly relationship with my neighbors before I brought up the bees. I think it's important for people to maintain healthy, communicative relationships with the people in their neighborhoods. As a culture, I think it's safe to say that we've lost touch with the importance of community, and I think it is wise to work on having a better rapport with your neighbors *before* you start asking them to accept your seemingly nutty backyard hobbies. It's also just really nice to be able to lend a hand to someone if he or she needs it and to get the same treatment in return. It's an arrangement that makes sense to me. If your neighbors know you and know you aren't certifiably insane, they might listen to reason when you tell them that honeybees are a great addition to the neighborhood. Imagine if your very first conversation with your neighbors was to tell them you were putting 60,000 stinging insects in the backyard. It probably wouldn't go over very well. There has to be trust there, and trust is something a person has to work at; so start getting to know your neighbors before you order your bees!

Now, all of this isn't to say that I think a beekeeper should kiss up to everyone on the block in order to get all neighbors on board. I am talking about getting to know the folks in closest proximity to you. For everyone else, I think "out of sight, out of mind" is the adage that I would apply to urban beekeeping. As I mentioned before, rooftops are a great spot for bees in cities thanks both to sun exposure and the fact that they're out of the way of humans and other animals. As an added benefit, they are also the most inconspicuous. I suggest that all beekeepers take measures to keep the majority of passers-by totally oblivious to the presence of their beehives. Set your apiary site up to be as safe and nonthreatening as possible, and strive for invisibility. In most cases, no one will even know they are nearby until you tell them. And once somebody realizes they have been living near bees all along, they are more likely to accept your choice of hobby.

If it's legal to keep bees where you are and you, as a responsible beekeeper, take measures to keep well-maintained hives in a location that doesn't inconvenience anyone, there's really no reason why you can't have your honey and eat it, too. Just be honest, responsible, and neighborly and even the most skeptical neighbors will perk up a little when gifts of honey show up at their door each season.

Of course, some people just don't like bees, or they worry about severe allergy-related reactions, which are rare. What becomes important at that point is that you weigh the facts and come to a conclusion about whether or not you want to move forward in spite of one person's feelings or opinion. (See page 63 for more facts about allergies and reactions.) If facts and warm fuzzy feelings don't convert that one nay-saying neighbor, then use your best judgment to do the right thing. If you decide to go forward with the plan to keep them anyway, be the most responsible beekeeper you can be, following ordinances and laws and being conscientious about placement, and be sure to slip a jar of honey to your neighbors after your first harvest to help smooth things out if they get a little tense.

Or, you can look for another place. If you really want to be a beekeeper, you don't want to be kept from your dream because a few people think having bees in the neighborhood is a bad idea. Place an ad on Craigslist, connect with your local beekeeper's association, or just ask friends if any of them know of potential sites nearby where you could place a couple of boxes of cuddly honeybees. Put yourself out there, and, in time, a suitable spot will most assuredly pop up. Most of the hives I maintain are in places other than where I live. You'll find people out there who will help you; you just need to ask.

But, if you do have a few neighbors who seem as if they might actively oppose beekeeping or who have actually gone forward to do so, try not to take it personally. Most people have little to no experience with bees beyond being stung by some insect at some point in their lives. Most people can't tell the difference between a wasp, a hornet, and a honeybee, so it's their association with insects that's motivating them. Try to be understanding and do what you can to treat them with consideration, but be persistent about proving to them that bees are our friends, are tremendous contributors to our local ecology, and are not a threat to anyone's health.

GETTING STARTED

Once you have worked out the details of *where* you will keep your hives, now you've got a whole other set of variables to consider. You still have to decide what sort of starter bees you'd like to buy and what kind of hive you'd like to keep. After that, you need to get the right tools and equipment to maintain and build the hive, and then you have to

WHAT TO SAY TO YOUR NEIGHBORS

Bee stings hurt. It's easy to see why many people assume that they're going to die when they get stung by a bee. And while swelling on your face or neck can potentially be dangerous, most people have a very, very low risk of serious reaction. Here are a few statistics to think about when you are talking to others about the perceived dangers of beekeeping: According to the Asthma and Allergy Foundation of America, one hundred Americans on average die annually from insect stings and bites. This includes, wasps, hornets, bumblebees, honeybees, and venomous spiders. You can imagine that a very small number comes from any one species alone. While loss of life in any manner is tragic, it is important to maintain perspective. For example: More than two hundred Americans die each year from food allergies. And four hundred or more lose their lives because of anaphylaxis caused by drugs meant to heal them. So, you could argue that honeybees are significantly less dangerous to human well-being than the food we eat or the medicine we take to get well.

And the fact is that bees already live with us, even in a city. Whether kept by us or left alone in the wild, they are everywhere; it's just that we usually take no notice of them. Next time you are at a park or see a planted flowerbed on the street, consider not only the honeybee but also other wild pollinators you will likely see there, drifting from flower to flower. Have a look; you may even spot a few right now. Take a moment to watch them work. If you are feeling brave, take your finger and reach out to one while it is working. Get as close to touching its body as you can. If you manage to actually touch it at all—a challenge in itself, since they never seem to stop moving—it'll likely just fly away to avoid being bothered any further. Pollinators at work aren't looking for trouble. They are looking for food.

Intent on finding nutrient-rich nectar and pollen, the lives of bees do not belong to them. They live for their colony. The work they do benefits their family. As beekeepers, it's part of our job description to enlighten others to this simple fact: Bees are not so different from us. They live for one another, and they can't thrive without community. One bee alone doesn't stand a chance in the world. The same can be said for human beings. Gently educating apprehensive neighbors to this might just help to put their fears to rest.

Top-bar hives are inconspicuous and generally lighter than conventional hives.

buy new bees and get them settled in their new colony. This chapter teaches you how to get started.

HIVES FOR URBAN BEEKEEPERS

The earliest types of hives—other than the ones bees make themselves—were skeps and gums, the latter of which were so named because they used to be made from hollowed-out gum trees. These are the hives you see depicted in works of art throughout the ages—they're made of a basket or a hollowed-out log with wooden bars or boards at the top to which the bees fix their comb. While certainly romantic and visually striking, skeps and gums are illegal in much of the United States because they are often built with immovable parts that don't allow for inspection or treatment of the hive without causing serious destruction of the comb inside.

For this reason, this book will focus on the two modern hives, the Langstroth hive and the more simplistic top-bar hive, both of which are as easy to put together as Ikea furniture. I strongly suggest that any new urban beekeepers stick with these kinds of hives. They are the two best suited to urban beekeeping, and they are completely legal. Both contain movable parts that make inspections easy to do without destroying any comb, which is good news for the bees who spend a significant amount of energy making it. What's more, hives with movable parts give beekeepers a chance to better observe and learn about the colonies, the food sources available to their bees, and the productivity of queens. Without these glimpses into the workings of your hives, the fate of your bee colonies is often left to chance.

The Top-Bar Hive

Both inexpensive and easily constructed from found materials, a top-bar hive is the best choice for the beekeeper most interested in pollination and bee breeding. It is very popular in third-world countries, where many beekeepers lack access to the precision tools required to build other types of hives. Top-bar hives, or TBHs for short, are essentially a trapezoidal wooden box with removable wooden bars that fit across the width of the hive. The bees build a natural comb on the underside of each individual bar, using a beeswax-coated popsicle stick as a guide. A TBH can be made from salvaged wood (1x10s are the best for these) provided it is not diseased, rotten, or infested with pests.

Another benefit of a TBH is that you don't have to lift heavy boxes of honey or brood during an inspection. During the prime nectar flow, modern hive components can weigh 30 to 60 lb/ 14 to 27 kg. Top-bar hive inspections require only that you gently lift a bar of comb out, look it over thoroughly, and slide it over a notch so that you can move on to inspecting the next bar. As a result, it's also a perfect hive for those with physical limitations. You sometimes don't even have to use smoke to calm bees kept in TBHs, if they are known to be gentle bees, since the bulk of the bees are kept in the dark, undisturbed. (That being said, it's always a good idea to have a lit smoker on hand just in case; more on smoking later in this chapter.)

Keeping bees in top-bar hives has other great benefits, too. The combs in TBHs are unique in that the bees are able to build them freely and don't require the manipulative cell imprints of wax foundation, which is kind of like a printed wax guide that the bees follow. Bees that are allowed to naturally build

Combs in top-bar hives are built without the aid of a foundation.

their own comb will often produce more drones than they would in the somewhat constricting Langstroth hive, which uses foundation. Most people estimate that about 15 percent of a colony in foundationless hives is drones, compared to 5 to 8 percent of drones present in hives with a commercially made foundation. Over the years, it's been suggested by many beekeepers that since drones can't forage for honey and pollen they are a drain on food stores and thus expendable. That's why for many generations you could find only a "worker-cell imprinted foundation," which helps bees build combs but discourages development of drones.

But that's just wrong. Though drones don't forage, rear brood, or do any housekeeping, they do provide genetic material for newly queened colonies. Impregnating virgin queens is one of the most important duties performed by any honeybee, as it ensures the propagation of the species. Without enough drones for mating, queens would lack the ability to lay the fertilized eggs that would become worker bees. Furthermore, drones give off pheromones of their own that mingle with the other chemical signals in a colony, rounding out all of the communication necessary for highly functioning hive. Many natural-cell beekeepers report that their colonies are even more productive with a larger drone population than without. The reason: colony morale (which I'll talk about in chapter 4). And see page 71 for more on how natural comb can help restrict the spread of pests. This is not to say that natural comb is the magic bullet that will ensure that all of your bees will be healthy and productive, but it can be a step in the right direction.

Maintaining top-bar hives does have a few drawbacks. If honey production is your top priority, this hive is not constructed to allow for excess storage like the Langstroth hive is. You will harvest honey, but it will likely be a smaller amount, and the comb will probably have been used previously for brood. While comb that has been occupied by eggs and larva is totally harmless to consume and tastes the same, it won't be gleaming white like newly drawn wax. You also can't use a traditional mechanical extractor, also known as a centrifuge, to harvest honey from comb, as the two are incompatible structurally, and the more delicate combs can become dislodged from the top bar. To extract honey from natural comb, you have to crush and strain the wax. (See chapter 5 for more on methods of honey extraction.) But the benefit to this type of extraction is that you get another valuable product: beeswax. With the remnants of the

Keeping different types of hives can help beekeepers to narrow down what methods of management resonate with them.

comb, you can make lip balm, furniture polish, and body lotion. (If you like the natural comb aspect of TBHs but want to use an extractor, there are some TBH designs that allow for the use of Langstroth honey supers for extra storage space, so the honey could then be extracted in a centrifuge.)

Another important difference between the TBH and a Langstroth hive is how you handle combs. Since TBH combs are attached to the hive at just one point (the top bar), they are less stable and cannot be flipped horizontally to observe both sides of the comb. It is critical that you make a habit of viewing the frames by turning them vertically so that the comb doesn't snap from the top bar and tumble to the ground, sending angry bees up into the air. With Langstroth frames that have comb attached at all four sides, you can flip the frames easily top over bottom to view both sides with no fear of damaging your comb.

If you choose to start with a TBH, consider having a couple of extra miniature or "nucleus" versions of the hives on hand so that you can create new colonies from your original one when it gets crowded. This is called making a "split." Making splits can help prevent swarming, which can startle neighbors and become an inconvenience. In chapter 4, I will discuss how to perform a split with both top-bar hives and Langstroth hives. This is a great way for a beekeeper to get started in bee breeding and expanding one's bee yard without having to buy expensive packages. If you want to limit your bee yard to just a couple of

Some beekeepers camouflage their rooftop hives to avoid attracting unwanted attention.

hives, you can try selling off the nucleus, or "nucs" as they are called, to new beekeepers once you are certain that they are queenright and healthy.

Considering the ease of harvest, the low cost for initial start-up, and the lack of necessity for bulky equipment, I think top-bar hives are a great hive for a new urban beekeeper or hobbyist gardener to consider.

The Langstroth Hive

The Langstroth hive, which is the most widely used hive in the United States, is loved by many beekeepers because it is so easy to use. It's distinguished by its vertical construction, which consists of a series of stacked boxes filled with anywhere from eight to ten movable frames. The bees build comb within

the frames, and you can remove each one individually to inspect brood and harvest honey with little to no damage to the combs or the bees.

Lorenzo Lorraine Langstroth, a pastor and beekeeper in Massachusetts, invented this type of hive in the nineteenth century. After experimenting with different types of movable-frame hives, he noted that the bees seemed to honor a specific type of spacing in the hive. If a space was too small, they would fill it with propolis; too large and they would fill it with brace comb. He called this spacing "bee space," and it's about a ⅜-in/10-mm space in between combs. It's just about enough space for two bees to pass by one another back-to-back during their work in the hive. With this measurement, Langstroth designed the dimensions of the hive to minimize the amount of errant comb building and excessive propolizing, both of which can complicate inspections and disrupt the hive.

Langstroth hives are usually made with precision tools, making it easier to follow the rule of bee space. They consist of

several functional parts that, used in unison, create a structure that beekeepers can handle with ease. Bees don't seem to mind being kept in them either.

The Anatomy of the Langstroth Hive

All of the hive components listed here can be ordered preassembled from beekeeping suppliers, but it's often much cheaper to source them unassembled and put them together at home. I try to order any necessary woodenware a couple of months before I expect I'll need it. I also find that inviting friends over for pizza and beer in exchange for helping me assemble hives makes the process both fun and fast.

OUTER COVER
The outer cover helps protect the bees from rain and extreme sun. It is usually made of wood and protected with a layer of sheet metal that reflects sunlight and allows water to pool and evaporate. Usually outer covers are weighted down with a heavy rock or held down with ratchet straps so they don't blow off.

INNER COVER
Just below the outer cover is a thinner second cover, which helps protect and insulate the bees and prevent them from fusing the frames to the top of the hive. It's not essential, but it's often useful in conjunction with different methods of supplemental feeding.

HONEY SUPERS AND THE HIVE BODY
The main part of the hive consists of

A still life of the essential Langstroth setup, with protective gear and tools.

a series of vertically stacked boxes known as supers. Each super has eight to ten frames, depending on how big a Langstroth hive you built. The bottom two to four boxes constitute what is called the "brood nest." Bees typically raise their young in the bottom of the hive. When a beekeeper starts a new hive, the bees are introduced to just one super filled with frames, and as the bees begin to fill in all of those frames with wax comb, more supers and frames are added. Once three or four of these supers are filled with brood, the bees will typically begin building comb for honey

storage in any additional supers placed above the brood nest. A healthy hive can fill an additional two to four supers in a season, creating 80 to 120 lb/36 to 54 kg or more of harvestable honey.

FRAMES

In each level of the hive body and in each of the supers, you will find frames. These movable wooden components are the structures on which the bees will build their wax combs. The kind of hive I keep, a ten-frame Langstroth, has—you guessed it—ten frames. Each frame consists of a top bar, a bottom bar, and two side bars to hold it all together. In most cases, you'll need to assemble these with a few tiny carpentry tacks and some wood glue.

Once the frames are assembled, you'll need to insert either a sheet of wax foundation or a wax starter strip on the underside of the top bar. For new hives, I generally use a combination of both, placing some frames with small cell wax foundation and some frames with starter strips in the brood nest. This way, the bees can create ample drone comb in the foundationless frames. During periods of heavy mite infestation, this drone comb can be easily culled, effectively removing a large number of Varroa from the hive. For more on foundationless beekeeping as a means of integrated pest management, check out the sidebar on page 71.

SCREENED BOTTOM BOARD

Bottom boards function as . . . well, the bottom of the hive. Bees walk all over it, and it helps to keep predators or pests from getting inside. Bottom boards are available either solid or screened.

A screened bottom board can be helpful when you're assessing a hive's Varroa mite population. For this use, a sticky white board is added to the screened bottom board and will catch the mites, which regularly fall off the bees.

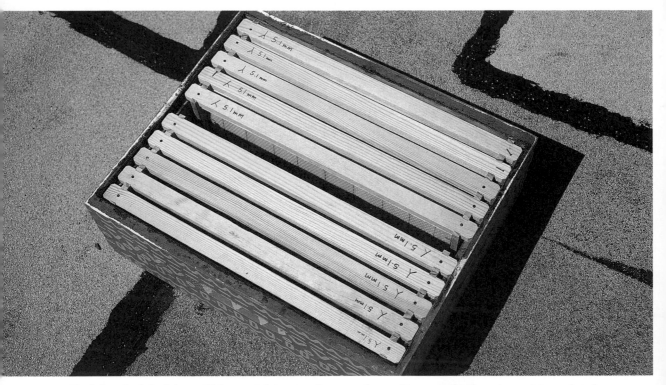

Each level of the hive will have eight or ten frames, depending on whether you opt to maintain eight- or ten-frame hives. I keep ten-frame Langstroth hives in my apiary.

MAKING WAX STARTER STRIPS

If you've decided that allowing your bees to make their own perfectly imperfect comb fits into how you'd like to manage your bees, well, I salute you. It's not always as straightforward as dropping in some cheat sheets for the bees to build off of, but it frees the bees to manage themselves a bit. Bees raised on natural wax comb tend to build slightly smaller worker cells over time and a greater number of drone cells—the ones that parasites like *Varroa destructor* (see page 103) prefer. Mites have a harder time breeding in the cramped worker cells and opt for the roomier ones that house drone larvae. As a result of allowing the bees to build their own comb, then, you may be able to avoid a potentially debilitating mite infestation of the hive.

Many beekeepers have adopted this management technique and have had great success. Fewer, if any, manmade remedies are used in natural cell hives. Folks from this camp generally try to create a situation for their bees where they can demonstrate resilience to the problems of disease and pests through instinctive behavior. I tend to fall into this category. All of my hives have a number of foundationless frames in them.

To successfully use foundationless frames, it is still helpful to provide a beeswax guide for comb building. This can be something as simple as a long strip of cardboard or several wooden tongue depressors glued into the wedge on the underside of the frame top bar. Once in place, brush a generous amount of melted beeswax over the strips from end to end. This hint of wax is all that is required to get the bees building comb on their own. This same method of making starter strips should be applied to top-bar hives.

If you choose to go foundationless, it's crucial that your hives are level. A tilted hive will yield a mass of fused-together frames that will give a new beekeeper serious anxiety. Cutting apart fused frames is a messy and disruptive practice, so level your hive and check in often while your bees get the hang of building combs without a guide.

These are the components of a frame: top bar, bottom bar, and two side bars. These can be used foundationless or a sheet of wax foundation can be used.

Hive Stand

Many people make or buy a hive stand to raise the hives up off the ground. Elevating the hive increases airflow, especially when your hive makes use of a screened bottom board. It also keeps the hive off damp ground, protects it from animals, and makes inspections easier. Be cautious about making a hive stand more than 8 to 10 in/20 to 25 cm high. Later in the season, when your hive is at its tallest, you may end up needing a stepladder to perform an inspection!

BEE GEAR

In addition to the woodenware you'll need for your hive, you'll also need to buy some tools and protective gear for yourself. With all of the variations of equipment available to beekeepers, navigating the beekeeping supply catalogs can be a pain. Since I like to keep things simple—and you might, too—I've come up with a list of the tools that I suggest all beekeepers have on hand before they get started.

Smoker

One of the most essential beekeeping tools is the smoker. Smokers are used to disrupt the flow of pheromones in the hive—especially the alarm pheromones given off by guard bees—by masking them with cool, opaque smoke. The smoke helps the bees stay docile, allowing the beekeeper to inspect a hive more easily. I recommend a smoker with a protective cage around its chamber; smokers can get really hot, and, from experience, I can tell you that it's easy to burn yourself on one.

Hive Tool

Hive tools come in many different styles, but they all do basically the same thing—they allow the beekeeper to pry apart frames that have been propolized together, and they are good for scraping burr and brace comb from places it shouldn't be. Hive tools, which are made of tempered metal, are meant to be strong. If you use tools you have laying around the house for working on the hives, they are bound to bend or break. My personal favorite is the J hook or "Maxant." With its little hook on the end, you can easily pry up frames from tight spaces.

Bee Brush

You use a bee brush when you need to gently remove bees from frames or out of harm's way. Unlike regular brushes, the bristles on brushes purchased from beekeeping suppliers are soft enough to avoid damaging the bees' fragile exoskeletons. More frugal beekeepers could use a large, clean feather, like a turkey feather. One farmer friend of mine in upstate New York actually uses a handful of sheep's fleece to brush away bees that have gotten between supers.

Hat and Veil

As far as protective gear or clothing is concerned, I really use only a hat and veil. I've been stung in the face before, and the results are rather unattractive. Bees, when miffed, will often aim for your head, so I try to keep that part protected. Bees also easily get tangled in long hair, so it's just a good idea to keep your head

A smoker (top), hive tool (above, left), and bee brush (above, right) are indispensable beekeeping gear.

covered. You can get veils that pull over other hats or buy hats with a built-in veil. I use one that you just pull over a hat of your choice because I like to look good while I am working with my bees! Just make sure that the hat you choose has a brim and no small holes in it through which a bee might be able to wander in.

Other Protective Gear

My policy is "wear what makes you comfortable." Protective gloves,

Wear whatever protective gear is needed to make you feel comfortable.

beekeeping jackets, or full suits and sleeve bands can minimize stings, though they are not always 100 percent effective. In my opinion, good technique—being careful and paying attention to your bees—is the best method of prevention, so to that end, you should wear whatever you need to feel at ease while working with the bees. If you're afraid or uncomfortable, you're going to do a poor job.

I personally wear lightweight, light-colored everyday clothes when I perform inspections. Light colors aren't as visible to bees, while dark colors make you a clear target. In the summertime, this usually means a T-shirt and denim overalls. I tie my hair back, put on my hat and veil, and work at an easy pace. I get stung pretty infrequently, but I won't honey-coat it—I do get stung. But the anticipation of a sting is worse than the sting itself, so I try not to stress about getting stung. That only makes it worse.

PROCURING BEES

Bees will not come to you in most cases, sad to say. This is especially true in urban environments where feral colonies are rare and oftentimes exterminated when found living in a fallen tree or other hollowed-out space. Instead, to get your colony going, you're going to have to find them. There are plenty of ways to get bees—you could buy a package of bees, buy an established nucleus hive, or buy an empty hive and try to attract a swarm—and all of the options have both benefits and drawbacks. These three methods are the best for urban beekeepers just starting out.

Package Bees

Many beekeepers agree that buying bees is the easiest way for newbies to start a colony. The recommended and most popular method is buying them in a "package" and then shaking the bees from that package into an unoccupied hive. A package will usually contain 3 lb/ 1.3 kg of honeybees (approximately ten thousand bees!), a punctured aluminum can containing enough sugar syrup to keep the bees fed while in transit, and a small cage where a newly mated queen and her attendants await their release. Since packages don't come with any comb or brood, they often come with fewer health problems, and since they're the easiest source of bees to get, they're the most popular. For that reason, you want to do everything you can to get your package sometime before the main nectar flow. Depending on your region, you will likely want to try to get your bees as early as you can. In areas with shorter growing seasons, like the Northeast, bees started by April generally have enough time to build up and put away before the winter, while in

A package of bees is 3 lb/1.3 kg of workers, a caged queen, and a food source packed for transit.

warmer areas you could push it a bit later. (I personally would not start any hives from packages any later than mid-May, but if you do get a late start, I have several methods of overwintering smaller hives that you can try at the end of the season, which I will cover in chapter 6.)

To get your bees by early spring, you should put in your order as soon as you can at the beginning of the new year. Beekeeping has become increasingly popular, and breeders and package operations really struggle to meet demand these days, so get your order in early and be prepared for a delay of a couple weeks just in case. Most of these package businesses are based in the southern United States, where early spring can be very rainy and cause a delay in production. Where buying packages are concerned, the early bird gets the bees. Note that your chances of getting bees sooner rather than later are increased if you put in a group order, so consider ordering with a bee club near you if you can.

A mini hive, known as a "nuc," is a great way to obtain bees.

And since most suppliers will not ship the bees to you, you will probably have to pick them up. Again, the easiest way to get packages is to coordinate with a local bee club, since they almost always do group orders with an apiary of their choice, often local, and the group makes the plans for transporting the bees. You can usually save money this way, too, as shipping costs are lower per package when buying in larger numbers.

Nucleus Hives

The nucleus hives, or nucs, are miniature hives made up of about three or four frames of brood in various stages of development, plus worker bees and a proven egg-laying queen. As a method of acquiring bees, purchasing nucs can be more expensive, with some going for as much as $100 to $200/£200 to £300 depending on the genetics of the queen. Nucs are made by splitting a strong colony; you take a few frames of brood and food to make a smaller nuc hive. Then, once the frames are in place, the bee supply company introduces an already mated queen or lets the bees in the hive rear their own. Once the queen is established and laying eggs, the nuc is ready for sale.

The benefit to getting bees this way is that the queen is already proven to be viable. She's been seen laying eggs for a week or more, and that gives you some confidence that your colony will have no trouble expanding into their new home. Even better, all you have to do to introduce the bees to your hive is move the frames from the nuc box to the empty hive, filling the rest of it with new frames for the bees to build on, which means you can get started a little later than you can with a package, too.

On the downside, nucs can often come with health problems. Beeswax acts as a sponge for pathogens, so when you get older brood combs like those in a nuc, they can have some brood disease. This is less of an issue when you buy from beekeepers who cull old wax comb regularly, so ask them if they do.

Swarms

The word *swarm* can terrify people, but swarms are actually bees at their most gentle, docile, and easy to handle. When in a swarm, they are literally a blob of bees. They're looking for a new home, and their mission is to stay together and protect the group. You can actually brush a swarm into a 5-gl/19-L bucket and gently take it home. Catching swarms is the best way for a slightly more-experienced beekeeper—or maybe a new beekeeper with an experienced friend—to get bees. Not only are swarms free, but swarms are the most natural way for bees to propagate, which appeals to beekeepers like me who prefer more natural methods of apiary management.

Swarming happens when a strong colony has begun to get a little cramped in the hive. Swarm season usually occurs between April and June here in the Northeast, but swarms can happen any time a colony gets crowded. The bees will divide into two groups, with the queen departing with more than half of the current workers. Once they've exited the hive en masse, the break-off cluster of bees lands someplace near the old location, while a few scout bees seek out a new home. Meanwhile the old hive consists of some queen cells, brood in various stages of development, and the other half (or less) of the workforce that has been left behind to care for the young bees and oversee day-to-day operations. Often swarms build up faster in a new hive than bees in packages or nucs can, which makes them prized additions to a bigger bee yard.

The goal for a beekeeper looking to create a new hive is to try and catch a swarm before it decides to crawl into the side of a house or a fallen tree. New beekeepers should really ask for help from more-experienced beekeepers when trying to capture a swarm. While the bees are generally quite docile at this point, they still require a level of handling that most new beekeepers might not be comfortable with or

Swarms are the most desirable way to start a hive due to the natural vigor of a swarmed colony.

prepared for. If you spot a swarm, call a local beekeeping club for backup and ask if a member would be willing to help you out by letting you start your first hive with the captured swarm.

In the early spring, more-experienced beekeepers will set up bait hives near found hives to encourage the natural clusters to set up shop in them. Attracting the bees to the bait hive will make it easier to transport them into a new hive to be cared for. Nuc boxes with frames of old drawn-out comb make a great bait hive strapped securely into a tree, but other less-expensive, lightweight options are available from beekeeping supply catalogs for new beekeepers who may not already have drawn comb to use to lure the bees.

The main idea is to make your swarm trap appealing to bees. A nuc box has an interior cavity that is just about the ideal size for a new colony. Put one or two frames of comb or foundation in and leave the rest empty. (Resist the urge to completely fill the nuc with frames, as it can cause scout bees to look around for a roomier place.) The entrance of the trap should be no larger than the size of a large coin, as scout bees tend to look for new homes that are easy to defend. And you'll also want to set up the hive about 15 ft/5 m off the ground, since bees prefer to be higher up to avoid animal invaders. (Some beekeepers strap their traps into tree branches or elevate them on poles similar to those one might use to elevate a birdhouse.) The old beeswax will attract some bees, but it's also recommended that you use a lure, such as a piece of paper towel saturated with lemongrass oil or a Nasanov pheromone lure that you can buy through beekeeping suppliers.

Be aware that it can take some time before a swarm trap works. It's also true that sometimes a swarm won't land where you want it to land. But if you set up your trap in the early spring when hives are in their biggest period of growth, you could catch a swarm in a matter of a couple weeks. Once you notice a lot of bee activity around the trap, you'll want to act quickly to confirm that you've caught a swarm and transport them into a permanent hive into which they can expand.

THE BIG DAY: BRINGING YOUR BEES HOME

By late March, if you plan to start a hive, the ultimate goal is to have everything assembled and ready to go before your bees arrive. Set up your hive in the designated place a couple of days beforehand. This will give you some time to move things around a little if you decide that the placement is not ideal for any reason.

Use a cinder block to weight down the empty hive; that will keep it from blowing away. And if you decide to paint your hives for longevity, make sure that the paint or varnish you use is completely dried and that the hive has been given a chance to air out. Hives giving off noxious fumes will most assuredly put off the bees, causing them to abscond the first chance they get. Choose light, reflective colors, and opt for paint designated for outdoor use.

Finally, if you are a procrastinator, make sure that at least the first two supers and frames are assembled and ready to go before the bees arrive—you can hammer the other ones together as you go. And it's also a good idea to have extra woodenware on hand. That way, if your bees grow too fast, you will have the extra supplies.

When you pick up your bees, you'll likely want to make plans to either drive or have a friend with a car drive you. If you live in a city, you often may depend on public transportation or taxis to get around. While packages and nucs

BEE INTRODUCTION CHECKLIST

When your bees arrive, you need to be prepared with the following items:

A spray bottle filled with water

A smoker and fuel

A hive tool

A hat and veil or any other protective gear

A bee brush

A hive top feeder

1 gl/4 L of sugar syrup

An "entrance reducer" available from beekeeping companies

Extra frames and supers

are usually sealed up pretty well for transport, I don't recommend taking tens of thousands of bees onto a bus or subway or even into a cab. To me, that just seems like a bad idea. If just one bee escapes, people can get really upset. You won't need to be suited up; you'll need only your spray bottle of water and a bee brush on hand so you can brush off any of the bees hanging on to the outside of the package before taking it into the car. If you miss a bee or two, don't worry. Just roll down the car windows, and the bees will fly out . . . or just let them continue to hang on to the box until you get them home. The bee just wants to stay with its kind: It's not likely to sting in this context. If it's a balmy day, feel free to give the package a nice misting of water with your spray bottle. Also important: While in transit, keep the package out of direct sunlight—the bees get hot really fast—and make sure the bees get plenty of air, too. Definitely don't put them in the trunk

unless you want some seriously unhappy bees when you open it up.

Once you get the bees to their new home, you should try to hive them right away. Some beekeepers wait until the evening to hive packages to decrease the likelihood of the bees leaving, but I've hived plenty of packages in the morning without incident. In my opinion, the sooner you can get the bees settled in, the better.

HIVING A PACKAGE

You've got your bees, you've got your hive in place, and you've got your gear at the ready. Hiving your first package of bees is a big moment for any new beekeeper. This method, when you get to the kernel of it, is simply shaking the bees from one box into another. I was really nervous my first time, and I was afraid that the bees were going to get really mad when I opened the box and put them in the hive. In my head, I pictured myself violently shaking them into their new home. But it was great—the bees all came out in this sort of fluid manner, and I got only one sting on my finger. This type of introduction can be easy to do with Langstroth hives or top-bar hives. And, without further ado, here's how you hive a package:

STEP 1: Remove the inner and outer covers from your hive. You should have only one super set up with ten frames filled with foundation or starter strips.

STEP 2: Take five of the frames out of the hive and set them aside. I usually take out the centermost frames, leaving a large gap in the middle.

STEP 3: Put on your protective gear, including the veil. Make sure to tie the veil on securely in order to prevent any gaps that will allow the bees to get inside.

STEP 4: Spray the bees thoroughly with sugar syrup through the screened sides of the package. Make sure to get both sides. This is done to immobilize the bees and get them focused on self-grooming rather than flying. It's important: The only time I've ever had difficulties during this

Packages contain cans of syrupy feed to sustain the bees during transit. Remove carefully to avoid crushing bees.

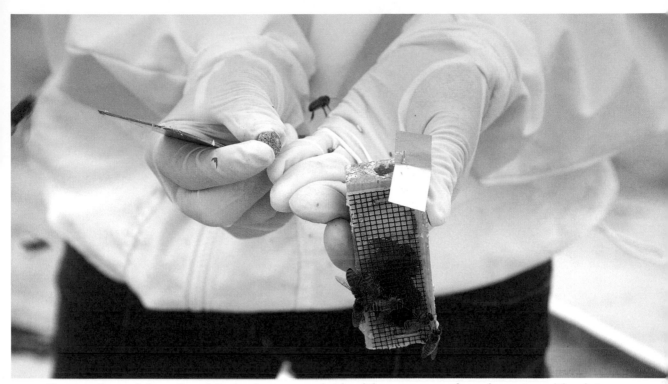

Remove the cork from the white fondant–capped side of the queen cage for a slower transition into the colony.

process was when I hived eight packages in a single day for local urban farms. My last package was for a hive that I had had to find a last-minute location for on top of someone's garage in Brooklyn. By that time, my veil was all sort of sticky and covered with spray, and I was tired, so I removed it. As I opened the package, I realized I had forgotten to spritz it down with sugar syrup. As the bees flowed out of the package, several of them did a U-turn and stuck themselves right into my face and neck. It wasn't awful because I had been stung before, but I had a triple-chin and lumpy head for the next day or so. Whenever you hive a package—whenever you're around your bees, actually—take your time and make sure you hit every step.

STEP 5: Give the package a firm whack straight down on the ground, just hard enough to knock the cluster to the bottom of the box. Don't whack it so hard that the box falls apart; do it just firmly enough to get all the bees that are clustered at the top of the package down to the bottom. Continue spritzing them with syrup until they are evenly and lightly misted.

STEP 6: Using your hive tool, pry the thin plywood tile out from the stapled corners of the box top. Keep this piece of wood on hand because it will be useful in the next step. Below the plywood tile, you will find the queen cage and the tin can filled with her sugar syrup.

Careful handling of the queen cage can help to ensure successful acceptance.

STEP 7: Slowly and gently remove the can. You may have to use a pocketknife or your hive tool to pop it out. You might be most shaky at this point, as some of the bees may start to try and squeeze through. Gently brush them out of the way as you wiggle the can through the gap in the top of the package. Once removed, slide the piece of plywood back into place, partially covering the opening again; this will keep the bees from flying out.

STEP 8: Remove the queen cage from the package. You will find some bees clinging to the cage, trying to feed her. Carefully shake them back into the box with the others before sliding the plywood over the entire opening of the package. Check the queen to make sure she is still alive and healthy. She should be about a third larger than any attendant bees she may have been packed with, and she may even be marked for easy identification. If your queen is dead, you will want to contact your bee supplier after completing the rest of the process. Do not discard the queen cage.

STEP 9: Examine the queen cage. It usually has two corked ends. One end will have white fondant, or candy, under the cork, which will be visible through the screen on the front of the cage. Remove this cork to give the other bees access to the candy. After removing the cork, poke a small hole through the fondant to get the bees started on releasing her. A small nail works fine; just work gently so you do not injure the queen. Over the next couple of days, the worker bees will slowly eat through the fondant to release the queen. This will give them time to accept her pheromones if they haven't already. This step is important, as the workers in the package are not born from the queen in the cage and will kill her if improperly introduced.

STEP 10: Wedge the queen cage, candy-side up, between the top bars of two frames in the hive. Make sure the queen cage is placed close to the center, where the bees will be warmer during the cool spring nights.

STEP 11: Remove the plywood tile from the package and carry the package to the hive. Invert it over the frames where you've just put the queen cage. Give the box a firm shake in an attempt to get as many of the bees out of it as you can. If a lot of the bees fly up into the air, gently set the package down and slowly walk away from the hive for a few moments until the bees go into

Hiving a package of bees can be daunting for a new beekeeper. Seek out help from other, more-established beekeepers if you feel uncomfortable doing it alone.

the hive to cluster with the queen. Once everything is calm, pick up the package and resume shaking out any remaining bees. Tilt the box to the side and rap it on the ground to get the bees balled up in a corner before dumping the remaining stragglers in.

STEP 12: Once most of the bees are out, gently begin putting the frames into the hive. There will be a cluster of live bees at the base of the hive, so be careful. A little wiggle of the frame as you put it into its place should get the bees to move out of the way. You probably won't be able to fit the last frame in, but that's okay. Take it inside and plan to have it with you for your first inspection. In fact, you'll always want to have an extra super and frames on hand for inspections in case your hive needs more room to expand. Slowly push the frames together toward the center of the hive so that the end bars are touching.

STEP 13: Put an "entrance reducer" in the hive on the smallest opening. This will allow your bees to more easily defend the hive against invaders.

STEP 14: Place a full hive top feeder on top of the newly introduced package. (See "Spring Feed for Bees," page 87.) The bees will need this food as they begin building comb and getting used to their new home. Place the outer cover above the feeder setup and weight it down with a cinder block or heavy rock. Do not open the hive for at least a week, when you'll be ready to do the first inspection. (See page 90.)

During the week that your bees are holed up in their new hive, they'll begin building comb for their soon-to-be queen to lay eggs in. Meanwhile, as some of the attendant bees eat their way out of the candy cork, other members of their adoptive family will be working their way in from the other side. With luck, when the bees finally meet their new queen, they will have accepted her pheromones. Once this acceptance takes place, the colony can begin its buildup.

TRANSFERRING A NUC TO A LANGSTROTH HIVE

If you've gotten a nucleus hive to start a hive with, you will likely want to move it to a Langstroth hive, since they are directly compatible. This makes for a super-easy transfer. I know some beekeepers who have broken apart the frames in a nuc, leaving the comb hanging from the top bar, and placed the resultant combs into a top-bar hive. Once you get more experienced, feel free to give that a try, but for beginners I suggest sticking with the more straightforward nuc-to-Langstroth transfer.

I recommend that when you get your nuc hive, simply set it down for a day or two in the location where your hive will be. Also open the entrance and

Handle frames from the outermost parts of the top bar. Move slowly and mindfully.

let the bees out so that they can fly around and get oriented. This will also give them a chance to calm down after being jostled about in transit. Just make sure you situate the nuc right where the full hive setup will be. After they've settled in for a couple of days here, move them into their new home. Here's how:

STEP 1: Light your smoker (see "Lighting a Smoker," page 95) and put on any protective gear that you feel that you need; a hat and veil are recommended.

STEP 2: Open up the empty Langstroth hive. You should have just enough frames with foundation or beeswax starter strips in the hive so that the addition of the frames in the nuc will fill it up. Remove any extra ones you might have and then push the new frames to the outer sides of the hive, leaving a gap in the center.

STEP 3: Gently puff some smoke through the ventilation screen and onto your nuc hive. Next, slowly lift the hive's lid, giving the bees a bit of smoke through the top, and then lower it again.

STEP 4: Using your hive tool, remove the frame closest to you. With both hands, pick up the frame and examine it, looking for brood. Try to locate the queen if you can. If you see eggs, but no queen, you can still assume that the colony is queenright. Move the frame to the new hive. Place it in the gap closest to you.

STEP 5: Repeat step 4 for each frame, being careful not to damage the queen if you locate her.

STEP 6: Once all of the frames are in the new hive, put in a feeder filled with spring feeding syrup. (See page 87.) Add your inner and outer covers and top them with a heavy rock or cinder block to prevent the wind from blowing the hive open.

STEP 7: Set up a water source, such as a bucket with some bricks in it, or a galvanized waterer used in chicken coops.

STEP 8: Come back in a week to perform your first inspection, to check the progress of your new hive, and to top off the feeders with fresh feed.

FEEDING NEW BEES

Believe it or not, when you first get your bees, you will have to feed them. This may seem strange, especially when there is plenty of food for them to forage outside in the spring, but when your bees first arrive, they will have little or no food stores with them. You will need to help them out a little until they have a chance to go out into the world, load up on nectar and pollen, and bring it back to the newly drawn combs that the house bees are building. The rule of thumb is to keep feeding young colonies until they stop taking the feed; it's usually not very long. Once they have lost interest in your sugar syrup, you can remove the feeder and let them forage for their own food.

A very simple top feeder is all you'll need, and you can add it during the first inspection or when you hive your bees. You can buy all manner of them, but I find that the best can be made with things that you will likely already have on hand. To make one all you need is:

A small nail

A hammer

Three or four 8-oz/240-ml canning jars with lids and rings, all freshly washed

4 cups/1 L of spring feed (See facing page)

An empty medium super

This is an inexpensive and efficient style of feeder.

SPRING FEED FOR BEES

(Makes approximately 1 gl /3.8 L)

11 cups/2.6 L water

5 lb/2.3 kg white sugar

In a stockpot, bring the water to a rolling boil. Turn off the burner. (This step is important because if the sugar caramelizes during this process, your feed can make your bees sick.)

Mix half of the sugar into the hot water, stirring until it is completely dissolved. Once the mixture is no longer cloudy, add the remaining sugar and stir until fully dissolved. Remove the pot from the burner and let cool.

Once fully cooled, you can add this mixture to your feeders or store it in the refrigerator for up to 2 weeks.

STEP 1: Using the small nail and hammer, puncture each lid with about a dozen holes spaced out all over it.

STEP 2: Fill the jars to the brim with spring feed.

STEP 3: When you are introducing new bees or during your first inspection, take the empty super and full jars to the hive.

STEP 4: Once the inspection or transfer to the new hive is complete, place the empty super directly on top of the open hive. Gently invert the feeder jars directly on top of the frames. Put the outer cover of the hive onto the empty super containing the feeders. Do not replace the hive's inner cover; just take it inside until it's time to remove the feeder jars.

STEP 5: Check your water source to make sure it is full.

CHAPTER 4:

So, You're a Beekeeper . . . Now What?

The first few months you have your bees are the prime time to get acquainted with them. The weather is nice, and they're usually in a good mood. The nectar flow is starting to kick in, and they're ready to go out there and start getting it—all of which means they're busy and they've got something very important to focus on. The hive itself is also smaller; it's less crowded and not yet gummed up with propolis or burr comb, making it easier to inspect and view. This is your best opportunity to get up close and personal with your hive—and you're going to learn the most about bees and beekeeping during the very first couple of months you keep them.

THE FIRST INSPECTION

Now that your bees are tucked safely into their new home, what do you do? Nothing, at least for a few days. You want to keep your mitts off the hive for about a week and let the bees get cozy. The workers will need to get acquainted with their new queen's pheromones, and work to eat away the candy capping that has her sealed in her cage. By the time she is released, they usually will have accepted her, and she will begin laying eggs in the comb they began building the day they were introduced.

Another key reason to just "let them bee" is that it allows for the colony to settle in and feel safe and comfortable enough to stay put. Bees will move to another location, or "abscond," if they feel like their current location is undesirable for any reason. Giving them peace and quiet while they get started means you have a better chance of your bees making the home you've given them more permanent.

But after five to seven days, you will want to go in for your first inspection. You'll know that the bees will not likely abscond if they are building comb and brood is present.

Once you open the hive, you will have about fifteen minutes to do what must be done during the inspection; after that the bees get antsy. In that time, you will want to open the hive; gently pull out the frames one by one; and look for evidence that your queen has been released, is accepted, and is laying

eggs in the wax cells that the bees have quickly built. With a Langstroth hive, the first thing you'll do is remove the queen cage and make sure she has been released. If she hasn't, remove the cork at the bottom of the cage or gently remove the screen, and let her walk out on her own and into the hive. Then close up the hive without doing an inspection and come back a week later to check the progress of the new queen. If the hive still has problems at that point, you'll want to read the section on "Dealing with Disaster" (page 97).

Inspection Basics

During a hive's first season, inspections are usually a breeze. The hive has fewer bees, so it's easy to find the queen, spot larvae, and move frames around without fear of squishing workers. As the first season progresses, however, the hive will get more populated, and the mood of the hive will change from day to day. Young hives are a great way for new beekeepers to get their bearings. I've said it to all of my students, and I'll say it here, too—working with a small, new colony is a dream. They're often gentle as lambs in the spring. They won't always be like that, so enjoy it while you can.

No matter what, you will likely be a little jumpy the first few times you open a beehive. That's totally normal. You've got to work at breaking the social conditioning that most of us have regarding bees. Do not be afraid to get stung. It's unavoidable. You should, of course, try to prevent making errors that would result in getting those stings, but if you are constantly worried about it, it will affect the way you handle the bees. I don't think bees can smell fear, per se, but jittery, unsure movements won't enhance your experience. You'll gradually get comfortable working your bees if you just follow some basic rules:

1. Approach the hive from the sides. Working near the entrance will ensure two things: (1) you will become a roadblock for foragers coming and going, and (2) guard bees at the entrance will notice you loitering and will get defensive pretty quickly.

2. Make sure your smoker is properly lit and try to avoid being heavy-handed with the smoking. A little bit goes a long way, and it becomes ineffective when you smoke them excessively. Use just a few puffs here and there when you first open the hive and as the bees begin congregating on the top bars. It'll send them back into the hive where they won't get in the way.

3. Move slowly and fluidly. Bees, like most critters, don't like spasmodic, abrupt motions. Think of an inspection like a slow dance with your bees. Gentle grace is a quality that I think bees appreciate.

4. Be mindful of where you place your hands. Most stings (for me, at least) occur on the hands. I dread finger stings—they hurt like the dickens!

5. Always take one or two frames out when inspecting a super and set it to the side, out of the sun. This makes moving the frames around within the hive much easier. Make sure you put them back in the order that you found them. Numbering the frames is a good idea if you are forgetful.

6. Slide frames into position. With a couple frames removed, you should be able to do this easily. Wiggle any frames that have bees between them. A few moments of this movement will motivate the bees to move out of the way. Do not cram frames in between one another. You could crush a lot of our little friends that way and can damage combs.

7. Limit inspections to fifteen minutes or less. If you are pressed for time, skip inspecting the honey stores (a peek between the frames can tell you a lot about the progress of comb building) and go straight for the brood nest to check for the presence of the queen and ascertain the overall health of the colony.

8. Consider what you smell like when you approach the hive. Bees are equipped with powerful scent receptors, so you don't want to show up smelling like a perfume shop or like you've been out all night drinking with your friends. Both are good ways to get noticed really fast by guard bees. Many beekeepers swear up and down that eating bananas before an inspection is a surefire way to get the bees acting defensively, as the scent of bananas is similar to the alarm pheromone given off by guard bees. I'm not big on bananas, so I've never tested the theory. In either case, being as inconspicuous as possible is always good policy when you are a beekeeper.

9. Time your inspection for when it's sunny and warm. A cloudy, cool day and the threat of rain and chilled brood can turn even the most gentle colony grumpy. If time dictates that you have to work with the bees when it's overcast, wear your hat and veil and keep it brief.

Inspecting a Langstroth Hive, Step by Step

If you are working with a Langstroth hive, your first inspection will go a little something like this:

STEP 1: On a sunny day, approach your hive from the side with a lit smoker and gently puff a bit of opaque smoke into the entrance. (To learn how to use a smoker, see "Lighting a Smoker," page 95.) Beforehand, double-check that your smoker hasn't flared up by testing it on the palm of your hand. The smoke should be no warmer than bath water. As a new beekeeper, I have inadvertently blasted flame into the entrance of a hive, which I can assure you the bees were none too pleased about.

STEP 2: Gently lift the outer cover just enough to puff some smoke

HIVE INSPECTION CHECKLIST

For inspections, you will need the following gear:

• Light-colored, clean, breathable clothing or a freshly laundered bee suit. Avoid perfumes, which the bees may be attracted to.

• A hat and veil and any other protective gear that you need to feel comfortable when working with the bees. Other pieces may include gloves, elastics for your sleeves and pant legs, etc.

• A hive tool, bee brush, and smoker.

• A book of matches or a lighter.

• An assortment of fodder for burning in your smoker. Sisal twine, untreated burlap, strips of old cotton T-shirts, leaves, pine needles, unprinted cardboard, dried grass, or black-and-white non-glossy newspaper are all suitable smoker fuels.

• A lidded jar for carrying away the bits of propolis and wax you will remove.

Always smoke the bees gently before entering the hive.

into the hive, and then set the cover down again for a moment, taking care to not crush any of the bees that get into the gap.

STEP 3: Wait a few seconds to let the smoke permeate the inside of the hive. Some beekeepers skip this step, but I don't. I'm in no rush, and you shouldn't be either.

STEP 4: Slowly remove the outer cover of the hive and set it upside down next to the hive. You'll use this as a place to rest supers as you work. Send a few puffs of smoke into the oval-shaped cutout in the center of the inner cover and remove any feeders.

STEP 5: Using your hive tool, pry the inner cover loose from each corner before slowly lifting it from the hive. The bees will likely have propolized it a bit. Try not to jostle the hive too much by roughly popping off the lid. This is the bees' first significant exposure to light, so it's best to move slowly. You'll have some

bees clinging to the underside of the inner cover. Gently set it to the side in a place where it won't blow over or become an obstacle.

STEP 6: Take a look between your frames. Are any combs built out? Get a quick count of how many appear to be drawn before you start prying apart the hive. You may be able to minimize the amount of damage to the hive if you take a second to observe and work smartly. If the topmost super seems to have no activity, use your hive tool to pry the whole thing off. Set it to the side. Be sure to set your supers atop the outer cover resting on the ground to avoid squishing any bees between the movable frames.

At this stage, use your hive tool to scrape off any errant comb being built on the tops of frames or any excessive propolis in between them. Place the bits of wax in the lidded jar that you will take with you when you leave your hive location, so as not to attract robber bees or predators.

STEP 7: Remove the outermost frame closest to you, using your hive tool to carefully pry up the top bar. Gingerly place your fingers on either side of the frame and slowly pull it upward.

STEP 8: Examine the frame. Is wax being built? If yes, is nectar being stored? The outermost frames often will be used for food storage. Once you've determined what is in that frame, set it to the side, out of direct sunlight.

Holding the frame with the sun to your back increases the visibility inside of the cells.

STEP 9: Using your hive tool, pry the next frame out. This one may have brood in it. Are there darker opaque, dry cappings located in the lower center of the frame? If so, you've got some brood near emergence. If the cells are not capped, take a glance inside. Is there nectar or pollen in the cells? Eggs and larvae? If you spot larvae, that means the queen has been there within the past week to ten days. Eggs indicate the queen has been there within the past three days. When you spot eggs, which can be hard to identify the first couple of inspections, note that each cell should hold only one egg, placed meticulously in the bottom center. Multiple eggs in cells can be a sign that you have some laying workers, which we will discuss later in this chapter. New queens do occasionally lay more than one egg per cell when they first get started, but a queen becomes more accurate after a day or so.

STEP 10: Once you've fully examined this frame, set it back into the hive in place of the first frame you removed. Try to avoid forcing it in, as you could accidentally injure the queen or crush workers, making your bees understandably defensive. Then carefully remove the third frame, examining it in a similar manner. Do you see eggs or larvae? How is the brood pattern? Is the queen taking advantage of all available space and skipping very few cells? Are the larvae a healthy pearly white color? Observe the workers. Are their wings smooth and aligned with one another? If the answers are "Yes," things are looking pretty good. If you answered "No" to any of these questions, read on to "Dealing with Disaster" (page 97).

STEP 11: Continue to go through, frame by frame, checking for evidence of the queen. As you check each frame, place it back into the hive in the space closest to the previously inspected frame. The idea is to create enough space to remove the frames without causing damage to the bees. Once you observe each frame in the super, gently slide the frames back to their original position. Any bees milling around between the frames can be coerced into moving by gently nudging the frame in a bouncing movement toward its final resting space. The last frame that you should place back into the hive will be the outermost frame that was closest to you. Repeat this process for each level of the hive until all frames are inspected, remembering to stack removed supers on top of the inverted outer cover. Do not place supers on their side, horizontally. You will crush many bees doing this, as the frames will be pulled down upon one another by gravity.

STEP 12: You've inspected the entire hive; now you'll have to put all of the supers back into place as they were before you started. At this point the bees will be fairly disrupted and maybe even a bit agitated, so you'll want to work swiftly but gently.

Use your bee brush to gently sweep the bees hanging around the edges of the supers back into the hive. Set the front corner of the super you are about to put down on top of the back corner of the super below it. Slide the super over slowly, brushing bees out of the way as you align them. Repeat until all supers are in their correct placement, with all edges lined up. Continue to watch where you place your hands, as there may be bees clinging to the outside of the hive.

STEP 13: Place the inner cover or full feeder back on the hive, followed by the outer cover. Replace any ratchet straps or weights on the outer cover.

STEP 14: Now you should clean up. Check to make sure you've picked up all bits of wax or propolis from around the hive to avoid attracting pests or predators. I like to keep a canning jar on hand to save the bits for making candles and propolis tincture (see chapter 7). Lastly, check to make sure that the bees' water source is both clean and full.

Remove the frame closest to you to free up space to handle the bee-covered frames.

STEP 15: Pat yourself on the back! You've just performed your first inspection.

Lighting a Smoker

Lighting your smoker is an easy task, if you follow the right process. Poorly lit smokers may go out mid-inspection due to poor airflow, or if they're not properly tamped, they may flare up, causing flames to shoot out of the mouth of the smoker. Both situations can create stress or a sense of panic in new beekeepers (not to mention the bees) during an inspection, which makes for a clumsy and potentially sting-filled experience. If you follow the ensuing steps, you should be able to produce a steady stream of opaque, cool smoke for the bees each time you visit, with little to no risk of flare-up. This will help ensure a more peaceful, stress-free inspection. You just want to take your time to get it lit and make sure it's done right. You don't want the smoker to go out when you're inspecting a hive; you are losing time that can be used to observe your bees.

STEP 1: Situate yourself someplace that is shielded from the wind. This will make the whole process infinitely easier. Nothing is more frustrating than going through an entire book of matches, trying unsuccessfully to light a smoker. On a rooftop, where it can be breezy, finding a windbreak is essential to success. Consider investing in a good-quality lighter if you expect to be inspecting hives often.

STEP 2: Start with a bundle of a quick-lighting, aggressively burning fuel such as paper or pine needles. Open the smoker and gently push a handful of this material into the chamber.

STEP 3: Tilt the smoker on its side before lighting a match and igniting the bundle of fuel. Once ignited, you can place the smoker upright.

STEP 4: Grab another handful of fuel and top off the chamber, puffing the bellows periodically. It's important at this stage to let air to flow freely in the chamber. Do not tamp down the fuel until it is completely ignited.

STEP 5: Once the fuel is burning steadily, begin adding slower-burning fuels such as burlap, twigs, and pine straw. I find adding a little bit at a time works best. As the fuel lights and begins smoking, add more, being careful not to snuff the smoker by packing it too densely. I sometimes use my hive tool to push the fuel into the chamber. Please be cautious! The smoker will be very hot at this point!

Varied natural matter makes a great smoker. Avoid possible chemical contaminants.

STEP 6: Give the bellows a few puffs every few seconds to really get the fuel smoldering.

STEP 7: Top the chamber one final time with another generous handful of slow fuel. Pack the fuel in so that it makes contact with the fuel already smoldering but not so hard as to put it out. Remember, fire needs oxygen, so you'll want some oxygen to be able to pass through the smoker. Close the lid securely, continually puffing the bellows.

Lukewarm, opaque smoke should pour from the spout. To test that the smoke is cool, just place the palm of your hand in front of the spout and give the bellows a puff or two. If the smoke feels no warmer than bathwater, your smoker is ready to go.

DEALING WITH DISASTER

Most of the time, you'll find that bees do a bang-up job of taking care of themselves. They find food on their own. They can often even rear a queen if the one they have has died or is performing poorly. They can manage some diseases on their own, without chemical treatment. And yet, sometimes, although you have done everything by the book, one of your colonies struggles to survive. In this section, I will discuss common problems and easy, natural solutions to each.

Bad Queens

One of the problems that beekeepers often confront is an unproductive or failing queen. It is important, especially for first-year hives, that queenlessness is tended to quickly. Colonies left without a queen for too long will have far fewer workers to go on foraging trips, which, in turn, means they will put away less food, have a smaller cluster, and stand less of a chance of making it through the lean months of winter. One clear indicator that your colony is no longer queenright is a lack of open cells of brood (especially eggs) during the spring or summer months. A productive queen can lay about 1,500 eggs a day. An egg remains in this stage of development for three days, so you should be able to spot thousands of open cells with freshly laid eggs at any given moment, especially if your colony has a strong matron at the helm. If you are not seeing eggs at all, take a second look. If you find the queen, but no eggs, you will want to replace her. This is called "requeening." You can go about this in a few ways, but I will focus on the two most straightforward methods: letting the bees raise their own queen and buying a new one.

Before you make a decision to do anything, you will want to confirm whether your hive is indeed without a queen or if the queen is in the hive and just performing poorly. To do this, you must do a thorough inspection of the hive, removing every frame and placing it in an empty super next to the hive until you've confirmed that the hive has no eggs, no larvae, and no queen. Many times during this process, novice beekeepers end up finding new brood after all, and a crisis is happily averted. But that's just one reason to check.

The other reason you want to go through your hive with a fine-tooth comb is to look for old queens. As a beekeeping friend once told me, "That old queen may not be able to lay eggs, but she may be a fighter." What that means is that you need to remove any competition to clear the way for the new queen. If you put a new queen bee in a colony that already has one, either the workers or the queen herself could seek out the newer leader and put a swift end to her. Or even if you put eggs in the colony and the bees raise a new queen to supersede the one you overlooked, that young queen could emerge and end up losing the battle to the existing matriarch. To avoid a whole lot of extra work, you need to make sure the hive really has no queens in it before you do anything further. So lock eyes on every single frame in that hive.

If, while going through the hive in question, you find a queen, either keep that frame separated or scoop the queen up into a small jar with a lid and continue on with your inspection. Once you've confirmed that no brood exists, and the queen is really a dud, decide on a course of action, and quickly. My preferred method of dealing with an insufficient queen is to remove her immediately, place her in a jar, and cover her with a bit of vodka. That will kill her quickly, and you can use the resultant "queen tincture" as a pheromone lure. But, usually, I will kill the queen

only when I'm able to act right away to replace her. In other words, don't kill the queen until you know what you're doing next and you're ready to do it. In some situations, timing is of the essence. As steward of your hives, you will need to decide if it makes more sense or you have enough time before winter to let the bees have at it on their own or buy a mated queen from a breeder and have it mailed to you. Then take one of the two actions outlined here.

OPTION #1:
FIXING A QUEEN THE EASY WAY

If you have more than one hive or have friends nearby who are beekeepers, an easy way to fix queenlessness is to give the colony eggs and larvae from a stronger hive. The bees will take the eggs and use them to rear a new queen, while the extra brood will help to keep the hive populated. This can be done quickly with very little waiting, but it can be risky as you will lose about three weeks of brood production during the time it takes for the queen to develop, emerge, mate, and then begin laying. Once the new queen begins laying, it will then take another twenty-one days before new workers (responsible for feeding brood and building comb) will emerge. The benefit of this sort of fix is that the bees are allowed to raise their own queen naturally, and she will be able to breed from the potentially diverse genetics in your community. I've had great bees come from naturally raised queens, so whenever possible, I opt for this method of dealing with queenlessness. Here's exactly how to do it:

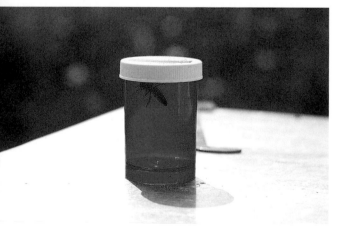

A recently removed, failed queen awaits her fate.

STEP 1: Using the same technique as you would during inspections, remove a couple of frames of brood from an established, queenright hive. I think two or three frames is ideal if the hive can spare it. At least one frame MUST contain a significant amount of eggs, some capped brood, and some larvae. Shake all of the worker bees off of the frame atop the hive, brushing any stragglers off with your bee brush. You will want to place new frames in the empty spaces left behind, before closing the hive up.

STEP 2: Place the frames of brood in a super or nuc box and cover them to keep them out of the direct sun. Brood can overheat and dry out quickly, so do not leave the frames out for more than a few minutes.

STEP 3: Smoke your queenless colony and open it up. Remove any supers above the brood nest and set them aside. Pull two or three frames with little to no activity from the brood nest. Oftentimes the outermost frames are not occupied with anything but food, so those would be good to remove.

STEP 4: Place the frames of eggs, larvae, and pupae into the center of the brood nest. The displaced frames can be moved upward into the honey supers or swapped out with undrawn frames if need be.

STEP 5: I suggest feeding your bees some pollen (real pollen, not a soy protein substitute) and either a frame of honey from a strong hive (which is always preferable) or some sugar syrup, so that they have some food available to them while things right themselves. You'd be surprised at how much this can help things along.

STEP 6: Gently close up the hive and wait. In a week or so, you will want to take a quick peek inside to check those frames. You should find several queen cells built onto the eggs that you transplanted. Right before emergence, I like to add another frame of capped brood and larvae from a strong hive just to give the colony some workers to help out with brood rearing and comb building, once the queen becomes viable.

OPTION #2:
INTRODUCING A MATED QUEEN

If you are in the midst of a big nectar flow or the colony itself is not large enough to effectively raise their own queen—lots of young bees are needed to feed the eggs and larvae royal jelly—you may have no choice but to order a queen from a breeder. This method of requeening is also useful in dealing with a "hot" or aggressive colony. A word of advice on buying-in queens: Stick with smaller open-mated operations whenever possible, and try to source them from local breeders if you can. Bees that are proven to thrive in your climate will be money better spent. Here's how to use a new-mated queen:

STEP 1: Order your queen from a local breeder who can ship her out promptly or will have one available for immediate pickup.

STEP 2: When your queen arrives, check the cage she's contained in to ensure that she is alive and vigorous. Moisten the screen so that her attendants can consume the water and feed her.

STEP 3: Open up the hive and remove the existing queen if you have not done so already. Dispatch the queen and place her in a sealed container so that you can remove her from the site. Do not leave the dead queen near the hive. Her pheromones will attract the bees to wherever her body lies.

STEP 4: Remove an outer frame in the brood nest, ideally one with no brood rearing taking place in it. Place the frame in an upper super or bring inside to replace at a later date. Move the frames outward from the center to create a gap. This is where the queen cage will be placed.

Don't want to kill the queen? Create a nuc for her to live out her days with some frames of brood and workers.

STEP 5: Remove only the cork of the queen cage covering the white candy capping. There is a cork at the other end that will release the queen immediately. Do not remove this! Using a short pin or tack, make a small hole in the center of the candy capping to stimulate the bees to start eating their way through. Be careful not to injure the queen during this step.

STEP 6: Place the queen cage in the center of the brood nest in the gap that you've created between the frames. Sandwich the cage between the frames in an "H" formation so that the screened side of the cage is fully exposed to the workers that will feed the queen and her attendants. Also be sure that the cage is candy-side up, so that any worker bees that potentially die inside the cage do not block the exit. For top-bar hives, you can position the queen cage sideways and facedown.

TIP: One way to tell whether a colony is really queenless or determine

A queen cage with attendants feeding her through the screen.

the mindset of your bees is to sit the cage containing the new queen on top of the open hive and watch. If workers begin chewing at the cage aggressively, that's a sure sign that either the queen is still in the hive, or they haven't yet begun to realize that she has been removed. If the workers instead flock to the cage and begin sticking their proboscises though the screen to feed the queen, that is a good indicator that the bees are aware of their queenlessness and ready to accept a new one.

Place the queen cage so that the screen and opening are accessible to the bees.

STEP 7: Close up your hive for three to four days to allow the bees to acclimate to the new queen's pheromones. On the fourth or fifth day, check to see if the queen has been released. If she has, remove the cage and replace the tenth frame. You can check to see if she has started laying eggs at this point, too, by removing a couple of the frames that were closest to the cage for examination.

If the queen hasn't been released, do it manually by very gently prying off the screen and letting her crawl into the hive. Be extremely careful during this process. Handling the queen is risky business, as any injury would result in rejection by her new colony. Also, it's not uncommon for a queen to fly out of the hand of an unsuspecting beekeeper who isn't quick about depositing her in the hive.

STEP 8: Close up the hive and resume your normal inspection schedule the following week to track the progress of your new queen.

Laying Workers

When a hive has been queenless for long enough that there is little or no open brood, some young worker bees will begin laying eggs in the hive. This is problematic, as the reproductively immature bees never mate and thus can only lay unfertilized eggs, or drones. The result of prolonged laying workers is a decline of the colony's health and strength. Egg-laying workers will occur infrequently if the beekeeper keeps a regular inspection schedule, catches queenlessness early, and remedies the situation with a new queen or open brood, as mentioned above.

You can identify laying workers in the hive by the presence of spotty egg laying with the telltale multiple eggs per cell. A queen will lay only one egg per cell, with the only exception being when a newly mated queen begins her egg-laying cycle. She will occasionally lay multiple eggs while she gets the hang of things. If you've not seen eggs in some time and upon further inspection see many sporadic cells filled with three to five eggs, you are likely dealing with laying workers.

To remedy a case of laying workers, most beekeeping books recommend a pretty messy solution that, in essence, requires the beekeeper to shake out the entire hive frame by frame 300 ft/100 m away from the hive location. The laying workers, who have never oriented to the hive location, become lost and cannot find their way back. In urban environments, this technique is impractical. Shaking an entire hive's worth of bees out into the open poses serious issues of safety. For that reason, I recommend that the beekeeper combine the queenless hive with a strong queenright colony. It's a risky move, but in most cases the workers in the strong colony will seek out the laying workers and remove them from the colony. After a couple of weeks, when signs of the laying workers have passed, you can split the hive into two again. (For instructions on how to combine a hive or split one, see pages 108 and 111.)

Robbing

When food becomes scarce in the honeybee world, our little winged friends become something like a gang of marauders. Bees are opportunists and will go after the most abundant sources of food in their territory. Sometimes that abundant source is another hive, and often it is a weakened one. Forager bees from one colony will invade another to steal from their food stores.

You will be able to detect this phenomenon, known as a "robbing frenzy," pretty easily. You'll spot worker bees fighting furiously at the entrance of the hive. Perhaps you'll even notice a large number of dead or dying bees on the ground. Sometimes a loud buzzing sound accompanies the battle. Especially severe frenzies will be seen as a cloud of angry bees fighting at the gates. This, as you can imagine, is not good for the hive that is being attacked and it is not good for neighborly relations for you. When you see robbing behavior, you'll need to act fast to put an end to it.

The way I've been taught to deal with it is pretty simple: Soak a bedsheet with water and completely cover the hive with it. This will block the invading bees and prevent you from having to get too close to your colony while it is on high alert. Keep the sheet on the hive until sunset, and then, once things have calmed down, remove the sheet and install an entrance reducer to its smallest opening. Close up any alternative exits your hive might have.

You'll want to check back the next day to make sure the robbing hasn't resumed. If all is clear, perform an inspection to make sure your colony hasn't been depleted of all of its stores. If there is little food stored in the hive, consider adding an internal feeder, described on page 86, and provide fresh pollen to help get the bees back on track. After a week, perform another inspection to make sure the colony is still queenright. Avoid external hive feeders like the boardman type, as they will further exacerbate robbing.

Hives that have been heavily robbed in the summer tend to need quite a bit of prewinter coddling in the form of supplemental feeding. So if you expect that your hive will need help through the season, decide now if you plan to feed, combine the hive with a stronger hive, or just let the bees tough it out.

Pests and Diseases

Honeybees, while generally pretty tough, are not impervious to illness or predators. Even the healthiest bees will at some point engage in a battle for survival against insect invaders, parasites, diseases, and animal marauders. As a beekeeper, you'll need to identify these threats to your honeybees' wellness and choose a course of action.

Pest and disease management are topics that can really pit beekeepers against one another. You'd be surprised at how differently many of us feel about the prospect of treating our bees for any illnesses. I expect that my perspective on matters will certainly anger some conventional apiarists because I believe in a slightly more hands-off approach to beekeeping, and my practices in dealing with disease and pests reflect that to some degree. This is not to say that I am willfully negligent with my hives, but I understand that like all living creatures, bees have the capacity to adapt and thrive in spite of external pressure if they are given the chance. The impulsive decision to treat symptoms with foreign, often toxic, inputs is not all that different than the human impulse to take medicine when feeling ill. Admittedly medicinal intervention is needed sometimes, but we are equipped with biological functions to recover from many things without them. As we become more dependent on medicine to cure us, the pathogens that cause the illness become resistant against these medications, and we either have to take more or upgrade to something stronger. It's a cycle that I make a conscientious effort to avoid personally, and I try to do the same for my bees.

Luckily honeybee health is definitely less complicated than human health. Fewer known ailments attack *Apis mellifera*, which makes it easier to play Dr. Beekeeper with them.

PESTS

Mites

Varroa mites, also known as *Varroa destructor*, are a parasite that was introduced to the European honeybee in the 1980s. The mites, which feed on the hemolymph of the bee, have reached a kind of stasis with *Apis cerana*, or the Asiatic honeybee, arguably due to the fact that many of the colonies are left to build wax comb without the guide of wax foundation. In the hive of the Asiatic honeybee, the mites prefer to reproduce in the significantly larger drone cells. As a result, the unmolested workers remain fairly healthy and are able to groom many of the mites out of the hive.

In the hive of *Apis mellifera*, however, mites became more of an issue. With this bee, beekeepers had adopted a different method of management in an attempt to boost productivity. In using foundation in the hives, beekeepers encouraged workers to create cells that were nearly a third larger in size. This, in turn, allowed the mites to breed in both the larger cells of the drones, as they do in *Apis cerana*, and also in those of the workers. As a result, the mites feed off of and reproduce in larger quantities on the worker bees, transmitting disease and, in many cases, killing the bees that are responsible for keeping the hive in working condition. Under the stress caused by these infestations, many colonies die.

While there are treatments on the market for managing Varroa, I stick to an integrated pest management, or IPM. This passive method of management gives the bees the opportunity to manage maladies themselves and develop some natural resistance in the process. Using small cell foundation in conjunction with several foundationless frames in the brood nest can help restrict

A Varroa mite feeding on a young worker bee.

Varroa to drone cells. Using screened bottom boards, culling the drone brood seasonally, and combining weak hives with stronger ones are other methods of management that can help keep you from buying things that you ultimately may not feel very good about putting in your bees' home.

Avoiding treatment, which I personally think is taking the long-term view of bee management, may not be right for everyone. You may have only one hive that you desperately want to keep alive, even as it struggles. If you feel as though using treatment is a route you'd like to explore, check out some of the references on page 169 to find information about conventional treatment of bee maladies. As someone who doesn't use them, I cannot speak to their effectiveness, so I will not pretend to.

Tracheal Mites

Another honeybee parasite, the tracheal mite, cannot be detected by the naked eye. This microscopic critter enters through the spiracles of the bee and latches on to the tissue in the trachea to feed, weakening the bee significantly. Laboratory testing is required to confirm the presence of tracheal mites in your colony, but symptoms can indicate a possible infestation. Bees with "K-wing," a separation of the fore and hind wings, as well as numerous worker bees scurrying away from the hive, can be signs of tracheal mite presence. These do not confirm an infestation, however, because the disease Nosema carries the same symptoms.

To sidestep tracheal mite issues, many

beekeepers select bee breeds with a known resistance to the parasite. Russians, Carniolans, and Buckfast bees all show some imperviousness to tracheal mites.

Wax Moths

Most often, infestations of the dreaded wax moth involve stored supers. This dusky-colored moth lays its eggs in unoccupied hive components. When the eggs hatch, little larvae spin a messy web, eating the precious beeswax as they grow. They then multiply, and the cycle continues, creating more damage. I cannot stress enough the importance of properly storing your gear. All wood-enware containing beeswax should be kept in airtight containers or double bagged. If you happen to spot wax moths on frames, freeze the frames for forty-eight hours before stowing them for the season. If you don't follow proper precautions to minimize wax moth infestations, you'll open up your supers to see them reduced to a gross and webby mess with little grubs crawling all over.

In some cases, wax moths will become an issue for occupied hives. Weak colonies without the workforce needed to kick out these invaders are usually the first to succumb to infestation. Remove inactive supers from these hives, and, if all else fails, consider combining the hive with a stronger hive with ample workers to give these unwanted bugs the old heave-ho.

Small Hive Beetle

Not entirely unlike the wax moth, the small hive beetle can overtake weak

An adult wax moth and the webbing left behind by its developing young.

colonies quickly. Once the adult beetle enters the hive, it will lay copious amounts of eggs in the cracks of the woodenware. It's not the adults or the eggs that wreak havoc; it's the larvae. With a voracious appetite, they tunnel through the honey supers, dining and defecating in the honey, effectively ruining it.

This shiny brown insect, which is only about the size of a hulled sunflower seed, is a serious problem in the southern United States but is often managed with inexpensive traps in chemical-free hives. This passive management tool is placed at the base of the hive. The trap consists of a mesh grid with openings large enough for beetles to fall through but too small for bees. Below the grid is a moat of olive or vegetable oil. The beetles fall in and effectively drown.

DISEASES
Sacbrood

A minor honeybee disease, sacbrood can be identified by dead larvae with darkened, shrunken heads and large, fluid-filled, sac-like bodies. It usually afflicts bees in early spring, but clears up on its own once the nectar flow begins.

Chalkbrood

A brood disease indicated by chalky, mummified larvae and pupae, chalkbrood is rare, but easy to spot. Worker bees in infected colonies will throw out the diseased brood. You will often find them in little piles outside of the hive entrance. Chalkbrood is a fungal disease and can often be prevented with good ventilation and the occasional culling of old brood combs.

American Foulbrood

One of the most dreaded brood diseases in the world of beekeeping, AFB is, for most colonies, a death sentence. Born from a prolific, spore-producing bacterium, the seeds of the organism live in the digestive tract of the larvae, diminishing its ability to feed. The brood then dies, and the matured bacteria release millions of spores, continuing the cycle.

AFB can be identified by dark, sunken-in brood cappings and dark, ropey, and melted larva. Many beekeepers say that a festering odor is also present.

If you suspect your hive is infected with AFB, contact your local extension agent to help you lay out a course of action.

European Foulbrood

Not to be confused with the more serious American foulbrood, this bacterial disease is responsible for larval death before the brood is capped. It can be identified by spotty brood patterns with yellowish-brown dead larvae in some cells. Disintegrated brood should still remain light in color, unlike American foulbrood, which results in dark, foul-smelling brood.

Nosema

Also known as "bee dysentery," the fungal malady Nosema can kill weak colonies as they cluster within the hive. It is identified by excessive fecal spotting on the entrance of and within the hive. Bees never defecate in their home unless something is wrong, so if you see bee poop within the hive early in the spring, you've likely got a case of Nosema on your hands.

Late-season feeding of sugar syrup and poor ventilation can exacerbate tolerable infestations, and the disease usually fades once the nectar begins to flow in the spring.

Deformed Wing Virus

Deformed wing virus is a disease transmitted by Varroa mites. The virus's effects are easy to identify: colonies infected with large numbers of mites will also carry a noticeable number of bees with stubby, malformed wings and pale, rounded abdomens. These bees are usually ejected from the hive and can be seen crawling around on the ground outside of the entrance. Implementing a good IPM regimen can help keep deformed wing virus and the mites that harbor it to manageable levels.

Swarming

One of the most misunderstood behaviors of honeybees is their inclination to swarm. It's a shame, since swarms are harmless and really cool to see. A honeybee swarm by definition is a reproductive division, a way for robust colonies to propagate the species by splitting into two, or sometimes three—or even four—new colonies. When bees swarm, the queen leaves the hive with more than half of the workers. The mass of bees with her exits the hive and lands someplace nearby, usually wherever the queen first lands.

Meanwhile, inside the old hive, several queens are nearing emergence from their cells alongside brood in different stages of development. One of the first queens to emerge will seek out the other queen cells, tearing into them and destroying the competition so that she—the first queen—inherits the role of matriarch. She will then inherit all of the duties of her mother, continuing the cycle of life within the original colony.

Swarming is an amazing occurrence to witness, though admittedly your neighbors might not be so pleased by the sight of a buzzing mound of bees covering the side of a house or amassed in a tree. So as city dwellers, we beekeepers have some responsibility to keep our bees from becoming a nuisance to our neighbors. Here are some easy ways you can help to prevent swarms without suppressing them completely.

ADDING SUPERS
Bees need space to continue to grow during the spring and summer months.

The general rule for expanding a ten-frame Langstroth hive is to add a new super once eight of the ten frames are drawn out with comb. There doesn't have to be anything in the comb, but if the frames are drawn, you should add a super. For this reason, you should always have an extra super and frames ready to insert whenever you perform an inspection.

Giving the hive added places for the house bees to build comb and the queen to deposit eggs can help to prevent swarming. Without that space, the foragers will start bringing back food quicker than the queen can lay eggs. As a result, the workers will begin backfilling the brood nest with nectar, effectively blocking her from using as many of the cells as she needs. They will also start building swarm cells containing developing queens along the bottom of frames in the brood nest and feeding the queen less in an attempt to slim her down. With frequent inspection of overwintered hives in the early spring, you can easily spot these clues.

OPENING UP THE BROOD NEST
Oftentimes beekeepers mistakenly assume that simply adding supers to a hive is enough to keep the bees from swarming. They will drop a box of frames down on a hive body packed with bees in April and, to their horror, a couple of weeks later the fire department is outside of their house trying to get their swarm out of a tree.

"Opening up the brood nest" can easily prevent a situation like this. It's a simple technique by which frames of brood are pulled up into the center of a new super.

SPLIT CHECKLIST

To perform a split on a crowded hive on the verge of swarming, you will need the following equipment:

- A nucleus hive or a medium super with four or five frames with foundation or starter strips, a bottom board, inner and outer covers, and a hive stand.

- A second super of new frames with foundation or starter strips.

- A feeder and serving of Spring Feed (page 87; in a pinch, a ziplock plastic bag filled with spring feed will work).

- A lit smoker, hat and veil, and any other protective gear you usually use.

- A friend who can serve as a second set of eyes for finding the queen. (Make sure your friend has the necessary protective gear, too!)

- An empty super set on the ground next to the crowded hive for placing inspected frames.

- An entrance reducer.

In their place, alternate empty frames between frames of brood. This allows for some of the younger house bees to start building wax on the empty frames, therefore alleviating the crowding in the brood nest, so the queen can then resume laying eggs. If you do this early enough, this often stops the swarming instinct.

"If they're building comb, they're staying home" is a beekeeping adage I've heard more than once, and it seems to be true. If you want your hive to feel as if it can grow without the need to divide, create more room in the brood nest until the bees have plateaued in terms of growth for the season.

WALK-AWAY SPLITS
Sometimes even opening up the brood nest is too little, too late. It happens to the best of us. A great source of pollen becomes available to the bees, and the hive explodes with new bees. One final step to prevent your colony from swarming is to simulate a swarm by splitting your hive into two or—in the case of a severely crowded apiary—three mini-hives. You may not want that many, but this can be a very temporary solution. To perform these splits properly, you should have a couple of wooden or cardboard nucs on hand or even an additional hive set up. This helps the process tremendously, so it's worth investing in these items as your bees overwinter. Have them ready to go once the weather breaks, just in case you need them. You may not need them that very first year, but when you or

one of your beekeeping friends end up needing one fast, I can assure you that you will be very happy to have the extra gear on hand.

To perform a split, approach the crowded hive and give the entrance a little smoke. Open the hive the same way you would during an inspection, puffing some smoke under the outer cover before removing it. Remove the inner cover and any inactive supers to get down to the brood nest. Give the bees a little extra smoke to keep them calm.

You'll notice that brood nest has a lot of bees in it—bees on top of bees. That is what beekeepers commonly refer to as "boiling over" with bees. It's the first indicator that the hive is crowded and really close to swarming. You may also notice that the bees seem really flighty and loud, or they might seem to be just milling about, inactive and aimless. This is also an indicator of a colony getting ready to swarm. Don't be frightened. The bees aren't usually any more aggressive at this point, but working in such tight quarters gives the beekeeper an increased opportunity for error. It can be a little jarring to see that many bees packed into one box, but just work confidently and easily, and you'll be fine. Working with the hive midday can help, as some forager bees will still be out collecting food, which will make the hive slightly less crowded. Remember: Good technique and mindful handling are the best prevention against getting stung.

Before you start the actual work of splitting the hive, set up your new hive in an appropriate location, facing southeast, and ensure that the hive is level. You will also want to set up an empty super next to your hive; after you inspect each frame and confirm it has no queen on it, you can place it in this super. We want to keep those frames isolated so that the queen doesn't backtrack over frames you've already checked out. She can be a sly one, that queen.

Using your hive tool, remove the frame closest to you and give it a good look on both sides. You are looking for the frame containing the queen. If you do not find her on the frame, set it into the empty super. Do not stop for queen cells, but note which frames you spot them on so they can be distributed properly before transferring them to the empty super. Go through each frame.

Once you find the queen, quickly transfer the frame she's on to the new hive and put it in the center. Move an empty frame either with foundation or a starter strip to either side of this comb to keep it protected from the sun and wind. Then go back to the inspected frames and pull four frames with worker brood at varying degrees of development. This means eggs, uncapped larvae, and capped pupae. Transfer these frames to the new hive, placing them on either side of the three centrally placed frames. Place the remaining new frames needed to fill the new hive on the outermost edges.

Now is the fun part: Go back and grab another frame of brood from the inspected super. Walk it over to the new hive and shake the bees into the hive with a swift downward movement but don't put the frame into the hive. You will want to do this with four or five frames; you just want the young worker bees on each frame to be transferred to the new hive to help build comb. The frames themselves will go back to the old hive so that this hive still continues to have some emerging worker bees.

Once you've shaken the frames into the new hive, add a feeder filled with spring feed, put in your entrance reducer, and close up the new hive so that the bees can begin to settle into their new home. Move back to the open hive and continue going through any of the remaining frames. Continue looking for any other queens. It's not entirely uncommon for two queens to be milling about that you happened to intercept just in the nick of time. Now you're going to select some of the larger queen cells that will be peppered all over the brood nest. Look for large, capped cells at the bottom of the frames. Remove any small, incomplete, or centrally built queen cells. I usually leave only one or two queen cells, just to make absolutely sure that the bees don't end up throwing a swarm anyway.

Once you've narrowed down your potential queens, put the frames back into the hive with some empty frames staggered between the frames of brood. This is effectively opening up the brood nest and giving the house bees some places to draw new comb. Once this is done, you can put an empty super on top of the hive and close them up. You should not have any problems with swarming this time around, but you should keep an eye on the brood nest and be sure to "open it up" whenever it looks to be crowded or lacks the presence of lots of very young brood.

If you decide that you would prefer to keep a limited number of hives and a new one puts you over that limit, you can transfer the five frames to a nuc and sell them to a local beekeeping club or recombine them in the fall. Just remember: If you want to keep this extra hive, you will need to feed the bees just as you would with any new hive. Keep feeding them until they stop taking it.

MOVING HIVES

It happens from time to time. Either due to neighbors who just can't stand having bees over their heads or something as simple as moving to a new neighborhood, a beekeeper will be required to relocate his or her hives. It's a prospect that seems frightening at first, moving an established colony with upward of forty thousand bees coming and going from the hive. But moving an apiary isn't as scary as it seems. It just requires planning, some heavy lifting, and lots of taping, stapling, and strapping things together.

You can't just up and move your hive anywhere though. You'll need to find a location several miles from the original location. This can be kind of tricky, but if you don't make the effort to break the bees' orientation to that spot, they will just keep returning to their old home. What does this mean? The forager bees will go back to the old hive location. When they get there, the hive will be gone but they will not know where to go. So, they'll stay put for a while until they find a nearby hive to drift to. The result is a congregation

COMBINING HIVES

If you've got a weak or queenless hive that is so far gone that it's doubtful any amount of fussing will keep it going, one option for dealing with it is to combine it with an established, queenright colony. In addition to your inspection gear, you will need:

- The weak colony with no queen
- One hardy, queenright colony
- A sheet of black-and-white newspaper
- A pocketknife

Combining hives is quite easy. Here's how you do it.

STEP 1: Inspect your weak, queenless colony. Confirm that there is no queen and minimize the size of the hive by removing any inactive frames.

STEP 2: While the newly condensed weak hive is open, approach and open the strong hive that you'll be combining with it. Use the same techniques as during a regular inspection, using your smoker and wearing necessary protective gear. Remove the outer and inner cover as well as any empty or inactive supers.

STEP 3: Lay the sheet of newspaper on the surface of the topmost super of the stronger hive, covering the top bars completely. Using your pocketknife, carefully poke a few holes in the paper to allow for airflow.

STEP 4: Starting with the lowest super first, transfer the weaker hive to the top of the strong hive. Place the supers on top of the sheet of newspaper. Place any of the weaker hive's upper supers on as well, if there are any.

STEP 5: Gently replace the inner cover, outer cover, and weights. Push the outer cover forward to keep the notch in the outer cover clear. This will provide air to the weak colony as well as an exit.

STEP 6: Refill the water source.

In a week, you should perform a routine inspection. You should find that the bees have chewed through the newspaper, which served as a buffer as the pheromones of the two colonies combined. The two colonies will have successfully become one!

of bees on the ground that your neighbors will probably not like very much. Hence, beekeepers have a rule to move the hive at least 3 miles/5 kilometres from the apiary site. The bees will be forced to reorient and will not return to the old hive location. Once the bees have reoriented away from the old site after a week, you can move them to the new location.

Of course, if the new location is already a few miles away, you only need to move the hive once. Before starting, be sure any unoccupied supers have been removed and harvest any excess honey to make the hive less heavy.

To move your hive, you'll need the following gear:

A lit smoker

Wire snips

A roll of fine wire screen

A roll of duct tape

Two to four 8-ft-/2.4-m-long ratchet straps

A heavy-duty staple gun

Two or three friends with strong backs and nerves of steel

A hand truck or dolly

Moving van or pickup truck

A hive net (optional; these are available from beekeeping suppliers to help keep bees that escape from the hive from getting loose)

Here's how you do it:

STEP 1: The evening before you plan to move your hive, you'll need to close it up for transport. Approach the hive and give it a little puff of smoke. Remove the outer cover and set it aside. Use the wire snips to cut an approximately 4-by-4-in/10-by-10-cm strip of wire screen and duct tape the edges securely over the opening in the inner cover. Tape the inner cover securely to the super below it.

STEP 2: Replace the outer cover. Run ratchet straps around the entirety of the hive so that one strap crisscrosses the other. The straps should hold together the following components, from top to bottom: outer cover, inner cover, supers, and bottom board. Tighten both straps so that the woodenware doesn't shift or vibrate.

STEP 3: Cut an approximately 36-by-12-in/90-by-30-cm strip of wire screen, to cover the entrance of the hive. The wire should be big enough to completely cover this opening. Honeybees can squeeze through pretty tiny gaps, so do not cut corners on this step. Once entrance activity has all but stopped for the day, place the strip of screen over it and begin stapling it into place. Give the bees a little smoke if they seem agitated by the noise or vibrations. It helps to have an extra set of hands to hold the screen in place. Be generous with the stapling. You don't want any bees escaping during transportation the next day.

STEP 4: Place duct tape around the edges of the screened entrance for added protection from escape-artist bees.

STEP 5: Early the following morning, gather your strong friends, hand truck, and van. If using a hive net, place it securely around the hive at this time. Slowly and gently lift the hive onto the hand truck, being mindful not to jostle the frames inside of the hive. Brush off any bees hanging on to the exterior of the hive. Place the hive in the vehicle and securely strap it in place to keep it from moving or tipping.

STEP 6: Once you are certain the hive is absolutely secure, you can transport the bees to their new location. If a couple of bees happen to get loose during transport, this is normal. You may have missed a few hangers-on. Just roll down the windows if you're in a van and let them out. If more than a half dozen or so bees accumulate, calmly pull over and check for gaps that the bees might be squeezing through. Tape up any suspicious-looking holes.

STEP 7: At the new apiary site, gently move the hive to its permanent location and place it on the hive stand, facing southeast. Leave the site for an hour—maybe go get coffee or lunch—to let the bees calm down a bit.

STEP 8: When you return, light up your smoker. Put on your hat and veil, too, as the bees may be quite agitated still. Remove the hive net if you used one. Puff some smoke into the entrance of the hive. Remove the ratchet straps and lift the outer cover, puffing smoke into the cutout on the inner cover.

STEP 9: Slowly remove the tape and screen from the inner cover. Bees may come out with stingers blazing, so just be prepared to take a step back. Keep your smoker handy.

STEP 10: Slowly remove the tape, staples, and screens from the entrance. Again, bees may be agitated at this point, so use your smoker if needed.

STEP 11: Place a weight or cinder block on top of outer cover and set up a water source for the bees.

And that's it! You just successfully moved your hive!

CHAPTER 5:

Show Me the Honey!

Honey is the edible pot of gold at the end of the rainbow for most beekeepers. You've taken your stings, and you've doted on those bees to the best of your ability. Now you're ready to eat some of the fruits of your bees' labor. I've been there: It's hard to watch the bees make beautiful food all season and not get a little bit antsy to try some yourself. I can assure you that the first time a beekeeper harvests honey is nothing short of a revelation, and the novelty of the honey harvest never wears off. Once you experience the smell of warm beeswax and see that amber comb glowing with the light of the sun, you will never grow immune to its beauty. Extracting the honey from the comb by hand is also a sensually sticky experience that further strengthens the bond between beekeeper and bees. For many apiarists, this is the profound moment when they fall madly and irreversibly in love with beekeeping.

HARVESTING

Honey harvesting can take place at any point in the summer once you've got some capped frames of honey in the supers above the brood nest. Many beekeepers will wait until the entire super is full to remove it. This means they'll get honey in one or two larger harvests, one in early summer and one in early fall, both times when resources are plentiful. Hobbyist beekeepers can also just take the occasional capped frame of honey out and enjoy it fresh out of the comb. My favorite way to eat honey is the soft summer comb crushed into a warm, crusty baguette.

Interestingly, my first beehive was probably the most productive hive I have ever had, thanks to a great queen. I probably got 80 lb/36 kg of honey off of that hive in its first season. I didn't even have to feed them in the winter as they had another 100 lb/45 kg to take them through. But that's rare. The first year you are lucky if you get just a few frames to take out; if you get a whole super, that's unusual. I would say realistically that if you don't feed your bees a lot of sugar, you can expect from 30 to 80 lb/14 to 36 kg of honey that first year. That is still an awful lot of honey for a hobbyist.

Before you harvest honey from your bees, you should always first determine that the hive is able to spare it. Do you have frames of full, capped honeycomb in the supers you have added atop the brood chamber? Is it early enough in the season that the bees will be able to replace the amount of honey you have removed from the hive? If so, you can

go one of two ways. The first option is harvesting individual frames here and there, extracting by hand, so you can then place the empty frames back into the hive after extraction. The bees will clean the spilled honey off of them and repair the wax comb quicker than you can say "sticky fingers." This method is suitable for the hobbyist beekeeper who would like to keep his or her honey jar full throughout the season and is not looking to turn a profit. I come from the school of smaller, more-frequent harvests. It's not so stressful on the bees, they always have access to the food they were intended to eat (instead of sugar syrup), and it's easy to do when you have only a couple hives.

The second method of extraction requires that you leave the full honey super on, without taking any of it until it is completely full and capped. Once it is full, you remove the heavy box of frames and replace it with an empty super for the bees to continue to fill until the end of the season. You'll then take those frames, uncap them, and extract the honey using a centrifuge all at once. This method works best for a beekeeper with more than just a few hives, as it helps to streamline the process of extracting, which can be quite inefficient to do by hand on a larger scale.

Beekeeping catalogs offer many tools that can help with the honey-harvesting process. If you plan to harvest individual frames periodically, however, the tools you have on hand for inspections will get the job done. A cardboard nuc is probably the only additional gear you might need to make it easier to carry multiple frames of honey from the rooftop. The rest of the extraction process can be done in a pretty scrappy way, utilizing basic kitchen utensils.

Harvesting by the Frame

To harvest honey by the frame, first, make sure you have the following tools on hand:

Protective gear

A well-lit smoker

A hive tool

A bee brush

A cardboard nuc with cover

Several frames with starter strips, foundation, or drawn comb

Then, follow these steps:

STEP 1: Approach the hive the same way you would when performing an inspection, in your protective gear and smoking the entrance first before removing the outer cover. Puff some smoke into the inner cover before removing it carefully. Perform an inspection as you normally would, using your hive tool, checking the progress of the hive, making sure the brood nest is open, and looking for signs that the queen is present. Once you've established that the hive is in good shape, you can start checking honey stores more thoroughly.

STEP 2: Go through the frames in the honey super, starting with the outermost frame closest to you. If you have multiple honey supers, you may have to remove the upper ones if you

find little activity or insufficient food stores in these frames. Inspect the frames one by one, checking for completely filled and capped honey cells.

STEP 3: Select a couple of the most completely capped frames of honey. Double-check these frames for the presence of eggs, larvae, or pupae. Any frames with brood present should stay in the hive. Be sure to replace them, keeping them positioned in the center of the super where it is warmer.

STEP 4: Holding firmly onto each end of a frame of honey, shake the bees off of the comb over the open hive with a sharp downward movement. Three or four good shakes should get most of the bees off of the comb. Afterward, step away from the hive and, using your bee brush, gently sweep off the remaining bees before putting the frame into the covered cardboard nuc box.

STEP 5: Replace the removed comb with an empty frame, positioning it between drawn-out combs to ensure the bees build the wax out straight.

STEP 6: Repeat steps 3, 4, and 5 for any other completely capped frames in the honey supers that you would like to harvest.

STEP 7: Close up the hive, placing the supers back in the order in which they were removed. Make sure all of the frames are properly spaced and that the supers are properly aligned. Place the inner cover and the outer cover on the hive and replace the weight or ratchet straps that keep the covers from being blown off.

STEP 8: Fill the waterers and pick up any bits of comb, propolis, or honey that have been spilled to prevent robbing. Carry the nuc with frames of honey off of the apiary site to be extracted.

Harvesting Using an Escape Board

One piece of equipment that I believe is absolutely worth buying if you plan to harvest whole supers of honey is called a "bee escape" or escape board. Some people use "fume boards" to separate the bees from the honey, but the escape board is a completely chemical-free tool for removing bees. It is nothing more than a board that is placed below the honey super that the beekeeper intends to harvest. In the evening when the temperature drops, the bees in the honey supers pass through a hole in the escape board to get back to the brood nest where their body mass is needed to keep brood warm. On the underside of the board is a screened maze that makes it difficult for the bees to get back up into the super. You should set up an escape board the day before you intend to harvest—just make sure the entrance notch under the inner cover is blocked to prevent robber bees—and the next morning your honey super should be nearly empty. Leaving the escape board on for more than a day, however, reduces the effectiveness of this tool, as the bees figure out how to get back in pretty quickly.

Here's how you do it:

STEP 1: After a thorough inspection, place your bee escape board under the full super you plan to harvest. It should be positioned with the cutout facing upward, and the screened "maze" facing downward. If you have an empty super above the full super, you can move it below the escape board so the bees can continue working on building out and filling those frames.

STEP 2: Close up the hive, placing the inner cover on top of the honey supers. Tape a piece of cardboard securely over the cutout on the inner cover and replace the outer cover.

STEP 3: Early the next morning, return to the hive. Light your smoker as you normally would. Puff some smoke into the entrance and remove the inner and outer covers. You'll probably find a few remaining bees crawling around inside of the honey super. Do not worry about them for now. Using your hive tool, carefully remove the honey super (it will be heavy!) and set it away from the hive.

STEP 4: Remove the bee escape and close up the hive again, removing the piece of cardboard from the inner cover and adding the outer cover.

STEP 5: Standing the honey super on end with the frames running vertically, puff smoke heavily into the super to force the remaining bees to evacuate. Once you've gotten most of the bees out, you can place the super into a large lawn-and-leaf bag and remove it from the apiary site for extraction.

EXTRACTING HONEY FROM THE COMBS

Your honey extraction process can be as easy or as complicated as you want it to be. It can be as simple as cutting the comb out of the frame or off the top bar and plating it with some fresh cheese, if the idea of eating wax comb appeals to you as it does to me. In fact, when I was selling my honey to restaurants in New York City, I would literally shake the bees off a frame and deliver it whole, so the restaurants could slice it from the wooden structure and serve it themselves. You can pop it into your mouth like a piece of candy or spread it on crusty bread, the texture of which helps break down the wax.

Some people, though, love the labor-intensive process of using an extractor or centrifuge. I can't say that I blame them. There is something methodical and meditative about spinning honey from a comb, just like the process of inspecting a hive frame by frame: You lose yourself in it. Sadly, for people like me who live in the city, an extractor is a rather large investment in apartment real estate—not to mention that the cheapest is about $250/£163—so I opt for a type of extraction called the "crush and strain" method (see following page). I think for first-year beekeepers to invest

Cut comb honey is the easiest to extract and enjoy fresh from the hive.

in an extractor is jumping the gun a little bit anyway. Instead, give this method of extraction a try if you end up with a couple of frames of honey early on. It's just as fun as using an extractor.

Harvesting Honey by Hand

CUT COMB HONEY

Cut comb honey is an absolutely no-nonsense method of harvesting your liquid gold. You simply take the frame into the kitchen and cut chunks of comb from the frame or top bar to the size of your choice. The key is to use wax starter strips in your frames, or special cut comb foundation that lacks the crimped wires that most commercially produced foundation contains. To produce cut comb honey, you will need just a few things that you probably already have in your kitchen:

> **Two cutting boards**
>
> **A frame of capped honey**
>
> **A clean piece of cardboard or construction paper cut to 2½ by 4½ in/6½ by 11 cm**
>
> **A sharp paring knife**
>
> **Newspaper**
>
> **A case of clean, dry 16-oz/ 475-ml canning jars with lids**
>
> **A spatula**

Once you have these items ready, follow these steps:

> **STEP 1:** Lay the two cutting boards on the counter side by side so that the entire frame can be laid across them. Place the frame horizontally on the cutting boards.
>
> **STEP 2:** Lay your piece of cardboard on the comb, with the corners flush against the corner where the bottom bar and right side bar meet.
>
> **STEP 3:** Using your paring knife, slowly cut around the piece of cardboard, leaving about ½ in/12 mm of wax on the underside of the frame's top bar. Leave the comb in its place until all of the pieces are cut.
>
> **STEP 4:** Move the piece of cardboard over to cut the next portion and repeat step 3. Repeat until the entire comb has been sliced into portions. There will likely be a sliver of comb left on the side. Trim that off and keep it as a little treat.
>
> **STEP 5:** Lift the frame slowly from the cutting boards by the top bar. The pieces of comb should slowly slip away from the frame back to the counter. Place the frame (with the ½-in/ 12-mm strip of comb still attached for the bees to use as a guide) in an empty super or nuc placed on top of some newspaper to avoid getting a sticky mess everywhere.
>
> **STEP 6:** Remove the lids from your jars and, using the spatula, lift each individual comb and lower it gently into a jar. With a damp cloth, wipe away any drips before replacing the lid. Repeat for each comb.

(Usually there is a considerable amount of honey left over on your cutting boards. I like to use a pastry

Crushing and straining takes no special equipment beyond what may already be in your kitchen.

scraper to get up every last bit, and add it to the jars of cut comb.

CRUSHED, STRAINED LIQUID HONEY

I jokingly like to call this "liquid honey for the lazy." This method of extracting honey is useful for those who don't particularly love the waxiness of eating comb honey or would like to add their harvest to their daily cup of tea. The only extra piece of equipment that might be helpful for this process is a bottling bucket system with sieves, which is available through most beekeeping suppliers. But an all-purpose kitchen strainer works just as well. Here's a complete list of all you'll need:

Two cutting boards

A frame of capped honey

A sharp paring knife

A spatula

Two large mixing bowls (with pouring spout if possible)

A wooden spoon

A strainer

A case of clean, dry 8-oz/240-ml canning jars with lids

Once you have these items ready, follow these steps:

STEP 1: Lay the two cutting boards on the counter side by side so that the entire frame can be laid across them. Place the frame horizontally on the cutting boards.

STEP 2: Using your paring knife, slowly cut the comb around the inside of the frame, leaving about ½ in/12 mm of wax for the bees to use as a guide when the comb is replaced.

STEP 3: Lift the frame from the comb and place it in an empty super or nuc box to keep the mess contained until the frames can be put back into the hive.

STEP 4: Slice the comb into pieces and, using your spatula, transfer the combs to one of the mixing bowls.

STEP 5: Using your wooden spoon, crush up the combs until most of the cells have been broken open. Resist the urge to crush everything until it resembles an emulsified paste; you end up with cloudy honey that way. If you miss any of the cells, you'll have a second chance to get them before the end of the process.

Chunk honey, a personal favorite, packs the most punch in terms of visual impact.

STEP 6: Set up your second mixing bowl with the strainer placed inside. (Or set up your bottling bucket, if you bought one, and the strainer.) Pour the mashed comb into the strainer, using your spatula to get every drop of honey out of the first mixing bowl.

STEP 7: This is the hardest part. Wait two or three hours for gravity to do the rest. I like to cover the strainer and bowl with a clean tea towel to keep bugs and random particulates out (but bottling buckets come with lids so you can just pop them on to keep your precious honey protected).

STEP 8: Check in halfway through the process. If honey is still trapped in some of the crushed comb, give it a gentle turn, breaking up the cells with any honey left in them.

STEP 9: Remove the strainer from the bowl and put it in the sink to be rinsed for wax rendering. (See "What to Do with Excess Wax," page 125.)

STEP 10: Fill your jars to the threading with honey by pouring it from the spout on your mixing bowl (or by using the honey gate on your bottling bucket). Little bits of wax are normal and will float to the top. (Allow the jars to sit out overnight and skim off the wax particles if you want a more pristine final product.)

STEP 11: Wipe off any drips with a clean damp cloth and screw the lid and ring on firmly.

STEP 12: Give your honey to someone special.

CHUNK HONEY

Can't decide if you want comb honey or liquid honey? Well, you can have the best of both worlds by making chunk honey. Chunk honey doesn't have the most sophisticated name, but it's easily one of the most lovely things to behold—a stick of honeycomb submerged in a jar of crystalline amber honey.

What you'll need:

> A few 2½-by-4½-in/6½-by-11-cm pieces of comb honey
>
> A clean, dry 16-oz/480-ml canning jar and lid for each piece of comb
>
> A spatula
>
> About half a comb's worth of extracted honey, ready to pour

Once you have these items ready, follow these steps:

STEP 1: Place each piece of comb into a jar using your spatula. Try to get the comb to stand on end to prevent gaps that might make filling the jar challenging.

STEP 2: Pour the extracted honey over the combs, making sure no air bubbles are trapped in the jar. Fill the jar up to the threading.

STEP 3: Wipe off any drips with a clean, damp cloth and screw the lid and ring on firmly.

STEP 4: Give your honey to someone special.

Using a Centrifuge

I had been beekeeping for about four seasons before I ever bothered buying a centrifuge. I just didn't have the space for it in my little Brooklyn apartment. Besides, I only had a few hives and tended to harvest smaller amounts, say ten frames from each hive, over the course of the season. The bees were left with more food, and I was spared the considerable investment for a piece of equipment I used only once or twice a year.

This is not to say that mechanical honey extractors aren't beneficial (and they are great fun to boot). Extracting using a centrifuge ensures that the wax combs that the bees work so hard to build can be reused by simply dropping the emptied frames back into the hive. The bees will polish the wax and start storing food in the cells as quick as a jiffy. As another benefit, when a beekeeper has numerous hives and has to harvest and extract the bulk of those honey supers in a straight shot, an extractor will always make easier work out of it. If you aren't ready for the

Extracted honey is strained to produce a clear final product.

commitment of buying a centrifuge but are curious to try one out, many bee clubs allow their members to borrow one or come to a group extraction day to take advantage of its usefulness.

The process of extracting via centrifuge has a quite few steps, but it's pretty easy to do. You'll need some extra tools, but most beekeeping suppliers will have extractor kits that are reasonably priced and take the guesswork out of which additional tools you need. For recommendations on where to buy an extractor, check out page 168.

If buying kits just isn't your thing, here is what you'll need:

> **An extractor of your choice (I suggest smaller, hand-crank models for the hobbyist urban beekeeper)**
>
> *(Continued...)*

A wooden pallet for securing the extractor

A tarp (optional)

One or more full supers of capped frames of honey

A basin or bin designated for uncapping frames (Rubbermaid totes work great for this)

A capping scratcher

A bottling bucket with honey gate and lid

Sieves for straining the honey

A couple of cases of jars and lids of your choice, washed and dried

Here's how you use an extractor:

STEP 1: Set up your extractor, making sure to mount it tightly to the wooden pallet. (I find that tilting the extractor forward slightly using a shim or an old book helps during the honey-bottling process.) Place the tarp on the floor to avoid tracking honey all over your home. Make sure all tools and equipment are clean and sanitary.

STEP 2: Remove your first frame of honey. Place it in the uncapping bin and begin removing the cappings from the frame using your capping scratcher. Using quick picking motions, slide the tines under the cappings before pricking them upward. Once all of the cells on that side are open, flip the frame over and repeat.

STEP 3: Place the frame in the centrifuge, into the frame baskets. The orientation of the frame will depend on the manufacturer's design, so consult your manual if it doesn't seem clear just by looking at it.

STEP 4: Repeat step 2 until the centrifuge is stocked with uncapped frames.

STEP 5: Put your hand to the crank and start spinning! I suggest starting slowly and increasing the speed gradually. Stand on the edge of the pallet to keep it from shaking or wobbling. You'll want to spin the frames for about a minute.

STEP 6: Open the centrifuge and check the frames to make sure one side of each is free of honey. If not, replace the frames, close the machine back up, and give it another minute of spinning. If the cells are clear, flip the frames and repeat step 5.

A frame of honey, ready to extract, should be at least 90 percent capped.

In order for the honey to flow, you must remove all cappings.

STEP 7: Place your bottling bucket in front of the extractor. Make sure the honey gate on the bucket (the opening through which the honey flows for bottling) is closed tightly to prevent leakage. Place a sieve in the bucket and open the honey gate on the extractor so any honey that accumulates can flow into the bottling bucket.

STEP 8: Repeat steps 2 through 6 until all of your frames have been extracted. All emptied supers with frames should be stored securely in lawn-and-leaf bags or put back into an active hive.

STEP 9: Transfer the remaining accumulated honey in the extractor to the bottling bucket. Allow the honey to strain into the bucket and then place the lid on top to allow the honey to rest overnight. This will allow air bubbles and small particles in the honey to float to the top. Thoroughly clean and dry all equipment for storage.

STEP 10: The next day, set your full bottling bucket on a sturdy counter or table. Have your jars at the ready. With jar in hand, slowly open the honey gate and fill the jar to the threading. Wipe off any drips with a clean, dry towel; place the lid tightly on the jar; and set aside. Repeat until bucket is emptied.

Store honey in the pantry or another cool, dark place. It will keep indefinitely.

WHAT TO DO WITH EXCESS WAX

You'll end up with quite a bit of beeswax after a day of extracting. There is no shortage of useful applications for this wonderful substance, so don't let it go to waste! Make candles (see page 161), lotions, and salves (see page 158). Polish your cutting boards. The list can go on.

But, even if you aren't quite sure what you'll use it for, you'll need to process the wax for future use. Leaving it laying around in its raw state is begging for annoying critter infestations. So, you'll want to wash and render your wax until you have a plan for using it. Additionally, you could sell blocks of your beeswax to folks who like to use it for some of the purposes mentioned.

The process to go from multicolored flecks of wax covered in honey to a usable material requires just a few steps, some patience, and a wax melter setup. The most eco-friendly way to melt wax is using a solar wax melter. Beekeeping supply companies sell these, but you can find a multitude of do-it-yourself plans in a simple Web search. Any beekeeper with the most remedial building skills can assemble one for less than $20/£15.

You can also render your wax easily in the oven, using some items you probably already have on hand. Just be warned, you'll need to designate these tools as "only for wax melting" after you use them once, as beeswax is notoriously challenging to remove from utensils. When rendering beeswax, exercise caution when setting the oven temperature and placing drip pans. Beeswax can be flammable at high temperatures.

To render your own wax, you will need:

> **Raw beeswax bits**
>
> **A colander**
>
> **Paper towels**
>
> **An aluminum roasting pan**
>
> **Cheese cloth, cut into an approximately 24-by-24-in/60-by-60-cm square**
>
> **Aluminum foil**
>
> **Two bricks or small stone pavers**
>
> **A glass or ceramic deep-dish pie plate**

Here's how you do it:

> **STEP 1:** Place the wax bits in the colander. Rinse the pieces thoroughly under lukewarm water, turning the

wax over to get any spots where honey has been trapped.

STEP 2: Dry the rinsed wax, spreading it on paper towels. Break up any large pieces by hand. Allow to air dry for two to three hours before turning the wax bits over to continue the drying process. Depending on the day's humidity, this can take anywhere from four hours to overnight.

STEP 3: Cut a 2- to 3-in/5- to 7-cm rectangular opening along the base of the sidewall of the aluminum roasting pan. This will allow the melted wax to trickle out of the pan and into the final drip pan.

STEP 4: Using the cheesecloth, bundle the dried wax scraps. Gently pack the wax into a loose ball. Tie off the bundle at the top and set inside the roasting pan.

STEP 5: Line the bottom of the oven with aluminum foil. Beeswax tends to burn and smoke heavily, so you'll want to avoid drips.

STEP 6: Position the oven rack at its lowest setting. Place the two bricks inside of the oven. You'll want them spaced out enough to serve as a base for the roasting pan, 8 to 12 in/ 20 to 30 cm apart. To create a slight incline, place one brick on its side and one flat.

STEP 7: Set the roasting pan, with the wax bundle inside, atop the bricks. The end of the pan with the cutout should be at the lowest position to allow the melted wax to leave the pan. Place the pie plate beneath this cutout to catch the melted wax. Double-check to make sure everything is stable and unlikely to tip over.

STEP 8: Set the oven temperature to 200°F/95°C. Allow the wax to slowly melt and drip into the pie plate. Check every 30 to 45 minutes to ensure the wax doesn't burn and the pie plate doesn't overflow. The entire bundle should be melted in 2 to 3 hours.

STEP 9: Once all the dripping from the aluminum pan stops, turn off the oven and allow everything to cool. To avoid spilling, do not remove the pie plate from the oven until the wax has set. Once completely cool, pop the wax out of the pie plate, cut it into pieces a size of your choosing, and store them in an airtight container until use.

HOW MUCH IS TOO MUCH?

In regard to the big honey payoff, I like to tell aspiring beekeepers that the first season of beekeeping gives you just a taste, but the second season is when you hit the jackpot. It's really critical that your bees end up with enough food to get through that first winter. Frames of honey do more than just feed the bees; they serve as insulation from the cold and help minimize drastic fluctuations in temperature. If your bees build up enough to fill a honey super or two in their first year, that is great. Feel free to take a couple of frames to sample but leave the lion's share to the bees during that precarious first winter. They need it more than we do.

Come late fall, each hive should have 60 to 80 lb/25 to 35 kg of honey stored away. This is the equivalent of four to five frames per super in the hive body. I usually leave hive bodies about four medium supers high to overwinter. I remove any extra supers, placing as many of the frames of honey as possible around the brood nest. I personally advise against taking an entire super in the first year. Just wait. If the bees make it through that first winter with lots of food stores left, there's no rule saying you can't take a couple of last season's frames from the hive in April to make room for new food stores. Honey doesn't spoil, remember? It's just as good after a winter in the hive as it was when the bees first made it.

Still, if your bees have made it through a winter, and you are ready to harvest an entire super, my suggestion is to do it earlier rather than later if you can. Give your bees some time to rebound on later nectar flows. I don't harvest honey from my bees any later than mid-September in my region, and when I do, I harvest lightly—no more than a few frames per hive. This gives the bees a chance to bring in some real food so I don't have to feed them sugar syrup in the fall. Sugar for honey seems like a rather unfair trade to me, and I'd rather my bees endure the winter on the food they worked all season for.

CHAPTER 6:

Wrapping It Up— End-of-Season Maintenance

If you assume that beekeepers get a break from thinking about their bees from the months of November to March, you are mistaken. Much work has to be done, even when the bees are clustered snugly in the hive. It pays to continue prepping—and learning—during the winter. Beekeeping isn't just about having your head buried in a beehive. First, you'll want to take some time to get your bees prepped for the months of inhospitable weather to come, and then you'll have some time to plan for the next season. Many of the activities I suggest in this chapter are things I do myself, so I can personally vouch for their usefulness.

WINTERIZING A HIVE

Once you've taken your last honey harvest for the season, you'll want to start preparing the hive to overwinter. In September or October, depending on your region, many of your bees' food sources will begin to freeze and die. Temperatures outside of the hive become too low for brood, so production slows to a crawl, and bees cluster together. Remember: Less food = less brood. Bees conserve during the winter, when times are lean.

There are many schools of thought regarding how much winterizing to do to hives—and much of that has to do with geographic location. Northern winters are notoriously blustery, so you may have to use insulation to add warmth to your hives. In the South, the milder weather makes that kind of winterizing unnecessary.

You will find minor differences in the degree of work that should go into winter preparations from location to location, but all beekeepers, no matter the climate, will take several necessary steps, including supplemental feeding, weatherproofing, preventative disease treatment, and deterring pests.

To Feed or Not to Feed?

In spite of your best attempts to help, some colonies will struggle to thrive. If your bees haven't put away sufficient stores for the winter, you may want to help them out by offering them supplemental food in the form of a fall sugar-syrup mixture. Some beekeepers will feed no matter what, especially after harvesting the bulk of the honey the bees have put away. Me? Well, I feel much better knowing the bees are in the hive all winter feasting on the beautiful honey they worked so hard to put away. I leave them with what they need to thrive and skip the sugar.

But suppose you didn't harvest any honey at all, and your bees just don't have the food stores they need to get by. What then? If feeding helps them to get by until the next season when they may have a better chance at flourishing, do it. Clearing out dead hives feels terrible, so feed your bees now if you are inclined.

You will want to start by feeding your bees a more-concentrated sugar syrup, or fall feed, as soon as the nights begin to get cooler. I use a 2:1 sugar-to-water mixture for several weeks. (See "Fall Feed for Bees," facing page.) I usually remove my feeders completely as soon as dropping night temperatures require that I turn on *my* heat for the winter. Bees need to have some time to let the water evaporate in the syrup before they cluster within the hive to keep warm. Nectar with excess water can cause more condensation, mold, and fungal activity in the hive, which can make the bees sick or kill them altogether. At some point, you'll have to cut the bees off, whether it seems they have enough food or not.

An added measure worth taking is to use the Mountain Camp method of

FALL FEED FOR BEES

(Makes approximately 3 qt/2.8 L)

6 cups/1.5 L water

5 lb/2.3 kg white sugar

In a stockpot, bring the water to a rolling boil. Then turn down the burner to its lowest flame.

Mix half of the sugar into the hot water, stirring until it is completely dissolved. Once the mixture is no longer cloudy, add the remaining sugar and stir until fully dissolved. Remove the pot from the burner and allow the mixture to cool.

Once cooled, you can add this mixture to your feeders or store it in the refrigerator for 2 weeks.

winter feeding. (See "Mountain Camp Feed for Bees," page 133.) This technique not only ensures that food is available to the bees all winter long, but it also helps to keep condensation from collecting within the hive.

Wrap It Up!

In colder regions, a beekeeper might want to contemplate what can be done to make sure his or her bees don't get too chilled during the winter. In some regions, it can get so cold in the winter that the cluster is unable to move to parts of the hive where excess honey is stored. In some of the most extreme cases, the bees end up starving. To prevent situations like this, many beekeepers opt to create an additional buffer between their bees and the elements. Some beekeepers wrap their hives in black roofing paper; others simply pile up leaves around the hives. I tend to opt for methods of winterizing that insulate the hive rather than attract heat to it during the day.

When a hive is insulated, temperature fluctuations inside tend to be more gradual. For instance, a hive that has black tar paper wrapped around it will heat up to a higher temperature during the day, perhaps to a degree that the

After wrapping a hive with insulation, install a mouse guard to keep out marauding rodents.

colony will break cluster. Once the sun begins to set, the temperature within the hive dips dramatically, and the colony may fail to cluster in time. A divided cluster has a more difficult time generating heat for itself. The result, in some cases, is colony death.

Some materials that work great for insulation are foil bubble wrap, bags of raked leaves, bales of straw stacked around the hive, and repurposed Styrofoam panels. Just make sure that whatever it is, it's secured around the hive, and it doesn't block the entrance in any way. Your bees will still venture out on mild days, so make sure their flight path is clear.

Weight It Down!

Winter wind gusts tend to pack more of a punch than their summer counterparts. With fewer leafy trees to serve as a windbreak during this time, you'll have to take extra measures to make sure that your hives don't get bowled over in a storm.

Ratchet straps, cinder blocks, or other heavy weights on the outer cover will protect the bees from gusts of wind.

Though bees tend to do a pretty good job stabilizing their home with ample propolizing in the fall, this doesn't mean that the hive itself can't be toppled. You will need to reinforce the good work that the bees have done. The parts of the hive that are most vulnerable to wind damage tend to be the inner and outer covers. I weight mine down with a cinder block, which works really well. Other beekeeping friends ratchet strap their beehives to their hive stands. Do whatever seems to make the most sense to you.

"No Vacancy"

For many critters, a beehive makes for very toasty and convenient accommodations. It's common for rodents and insects to find their way into a hive to keep warm. While it seems sort of sweet to imagine a bunch of tiny creatures cuddled up inside of a warm, honey-filled home, it's actually bad for the bees. Many of the animals that invade the hive during the winter consume many of the wax combs and tend to make a real mess. What's worse, they can disrupt the cluster, impeding the bees' ability to keep warm. That can spell doom for the colony.

In order to keep out unwanted hive crashers, you'll want to put a mouse guard on the entrance to each hive. These are available from beekeeping supply companies, and most are very easy to use and reuse each winter. They are usually just pieces of metal or plastic that a beekeeper can mount to the front of the hive. Bees can pass through little holes all over the guard, but mice

MOUNTAIN CAMP FEED FOR BEES

An easy, emergency food source for overwintered hives.

1 sheet of newspaper or an inner cover

1 spray bottle filled with water

1 empty medium super

5 lb/2.3 kg white sugar

Once you have these items ready, open up your newly condensed hive (about four medium supers high), using the same technique as you would during an inspection. Remove the outer and inner covers. On top of the brood nest, lay down your inner cover or piece of newspaper, covering the frames. Spritz it gently with water to weigh it down. Next, place the empty medium super in the hive. You will not need any frames for this, so leave them out.

Tear open the bag of sugar and gently dump it onto the newspaper. Some may spill into the hive, and that's okay. It won't hurt the bees. Use up the entire bag, spreading the sugar evenly over the frames. Then, just put the cover back on! The sugar will serve as a food source when liquid feed would freeze. On mild days, the bees will venture up and take what they need.

Many winters my bees don't even touch the sugar, but it does a fantastic job of wicking away any of the humidity the colony may create. In the spring, you will often find a hardened sheet of "candy" in the place where you layered on the sugar. For its humidity-regulating properties alone, I think the Mountain Camp method is worth employing.

cannot. Personally, being a bit of a penny pincher, I just opt for a ½-in/6-mm strip of hardware cloth, a heavy-duty metal mesh. It is cut into strips and stapled to the entrance. It does the same thing as a mouse guard, only it's made of a material that many people can find at their local hardware store if they don't already have some lying around.

WINTER TREATMENT OF PESTS AND DISEASES

You'll notice that this book is not heavy on advice about dealing with pests and disease. There's a reason, and it bears repeating. There's not much a bee colony cannot overcome if they have a strong queen, good ventilation, and access to good food and clean water. This is what makes for a strong colony. If you feel your bees aren't thriving, one of these pieces may be out of place. Try

THE CLUSTER

People are always asking me where the bees go in the winter. Do they die? Do they migrate to some tropical getaway with their avian counterparts? Turns out, they do pretty much the same thing we humans do during the colder months of the year: They hunker down and cuddle up.

In the winter, bees remind me of penguins in Antarctica. Once the outside temperatures drop below 60°F/16°C, they pull themselves together into a ball, or "cluster," to help conserve energy and warmth. They have the added ability to create warmth by vibrating the muscles located under their wings, creating friction and heat. When brood is present in the hive, the cluster will radiate temperatures near that of the human body.

While bunched up together in the hive, the bees migrate over frames of honey, eating slowly as they go. On days when the temperature outside breaks 50°F/10°C, they will take turns going on cleansing flights to avoid relieving themselves inside of their pristine home. They'll use these breaks from the cluster to remove dead bees from the hive and occasionally forage for water.

It is of the utmost importance not to disturb a bee colony when it is in a cluster over the winter. When disrupted, the colony can have difficulty pulling the group back together and the tragic end result could be a hive of frozen bees.

to correct it, and your bees will stand a pretty good chance of surviving. Try to bandage it by treating with chemicals, and you will continue to weaken your bees and the bees in the community who will inevitably share your colony's genetics during mating season. I'll get plenty of flack from beekeepers who have been keeping bees in the conventional, more input-dependent way for ages, but the treatment-free approach to managing bees is underappreciated. And it is one I stand by wholeheartedly.

So as winter comes and the majority of the beekeepers in your club are putting antifungal sugar syrup and chemical strips in their hives, you may want to instead offer your weaker hives some frames of honey from your stronger hives, along with some fresh pollen patties to help give them a boost near winter's end. Better yet, if you notice a colony struggling in July or August, do something about it then rather than wait. The bees will thrive if you give them the chance to.

I do not treat *any* of my hives. At all. Do I lose hives sometimes? Absolutely. It tends to be fewer hives than some of the beekeepers I know who utilize a more traditional approach to bees. My feeling is that if, with a good, sufficient supply of food, a quality queen, and good ventilation, these bees cannot survive, perhaps they were not meant to. Bees were never meant to be coddled and manipulated to the degree they have been for the past one hundred years. We can turn back the hands of time if we give the bees a chance to adapt to problems as they did for millions of years before we became their active stewards. It's a difficult choice to make to decide to let natural selection run its course. It's easy to question if such a choice teeters on the edge of willful negligence. I've weighed the options personally, and I feel that manipulating these creatures to an end that is beneficial in the short term has already been shown to have detrimental effects by beekeepers who are smarter and more experienced than I am. The hives on which I've used less-invasive management techniques—using equipment that more closely mimics their natural behavior, for example—have thrived. Ones that I maintained by "the book" tended to peter out after the second season due to heavy mite loads and mite-transmitted pathogens.

I do not treat my hives and probably will not ever do so. It's my goal to keep bees that aren't dependent on humans to survive. I want hardy bees, not wimpy bees. Any farmer worth his salt will expand his flock with only the strongest of his livestock. The ones that fail to thrive will often be culled to avoid harming the rest of the flock. With bees, nature tends to take on the brunt of the dirty work. If that sounds sensible to you instead of brutish and coldhearted, you may want to adopt a more passive approach to the management of your hives. If you'd rather be more involved in management of maladies, that's okay, too.

If you decide you do want to treat your hives for disease or infestation, see page 103 to help you to identify problems. You can purchase most remedies from the beekeeping supply companies listed on page 168. The section lists numerous forums where beekeepers from different schools of thought weigh in on what they think is the most effective. These are some of the exact forums that helped to inform the way I keep bees, and I expect they will do the same for you. Each beekeeper gains something different from the time spent with his or her bees and as a result, may have a different approach to management. I don't judge those who treat their bees; I simply choose not to. Only you can make the decision that's best for your colonies and you as a beekeeper.

ARE MY BEES ALIVE?

On mild, sunny days throughout the winter, you'll want to sneak a peek at your hives to see if there is any activity. When the temperature is above 50°F/10°C, honeybees will generally take the opportunity to stretch their wings and take care of some chores. If your colony is still hanging on, you'll see bees carrying out dead comrades, zipping out on a cleansing flight, or just milling about at the entrance a bit. This, of course, is a sure indicator that your bees are still kicking. But, just because you go up in January and your bees are humming in their hive, don't assume that they are in the clear. The months of February and March can be the toughest on honeybees in some regions.

You'll want to continue checking in on your bees. After a snowstorm, you'll want to visit your apiary to clear accumulated snow from the entrance so that air can flow freely and the bees can take a trip if they need to. On very warm days with temperatures exceeding 50°F/10°C, you may want to open the hive to see if the bees have emergency feed left. Give the hive a little lift. Does it feel really light? Are the bees accumulated near the top of the hive? If yes, you may need to concern yourself with supplemental feeding.

At this time of year, using syrup as a food source isn't recommended. The syrup will often freeze and bring the air temperature within the hive down with it. The easiest way to get food into the hive is to restock your Mountain Camp feed setup (see page 133) with more dry sugar and a spritz of water, and hope for the best. Keep checking on your bees' emergency feed every week, and if it gets low again, stock it up. As soon as the first trees come into bloom, your bees will jump all over them (if they have toughed out the winter), and you can lay off the sugar for the rest of the season.

WHAT TO DO WITH YOUR SEASONAL DOWNTIME

So, you may have had a great season and your bees show signs that they will survive over the winter. Now's the time to plan ahead for a successful spring. You may decide that you will start a couple more hives next year. Maybe you wish you had a more bee-attractive neighborhood or garden, with more diverse flora for your army of pollinators. Here's how to make the most out of your winter, so that your spring and summer are both stress-free and productive.

Expanding Your Apiary

If you had a great first season, you might end up deciding to start another hive or two. You'll need the winter to prepare. If your current hive is looking like it will survive the colder months, you might be able to just split that hive into two. Or you may decide to try to keep that colony intact so it gives a bumper crop of honey, meaning you'll need to find another way to acquire more bees. If you want to buy packages or nucs from your local bee club, you'll just need to order the extra woodenware required for the season and have it assembled and painted before the bees show up in the spring.

You could also choose to take a more natural approach and try your hand at catching a swarm in the spring (see page 77). Locally adapted, overwintered swarms are more valuable than gold for beekeepers. They tend to be very healthy with a great queen at the helm. They also tend to build up very quickly once they find a home. For this reason, beekeepers absolutely love swarms. In New York, beekeepers have been known to get into heated arguments over who claims the right to one! You might want to consider putting together a swarm-catching kit now, so that when you hear word of a swarm in your neighborhood, you can run off to nab it first.

Planting for Pollinators

City dwellers can help boost the amount of food available to insect pollinators by planting gardens that supply a little extra nectar and pollen where it's needed. Many of the plants that bees love happen to be useful to people too, either for medicinal and culinary uses or even just for cut flowers.

Herbs like basil, thyme, and oregano produce beautiful flowers that bees love, and they also happen to make wonderful teas! Mint, lavender, monarda (also known as "bee balm"), anise hyssop, rosemary, sage—in reality most aromatic herbs are appealing to bees. If you can use them at home, too, plant some extra. The bees will most certainly thank you in their own way. (You probably won't be able to influence the flavor of their honey much with strategic planting, however, unless you've got acres of these herbs available for them.)

There are also great varieties of flowers for cutting that you can plant to beautify your garden and provide food to pollinators: cosmos, zinnias, sunflowers, and asters are all tremendously valuable as pollinator attractors, and they also make for gorgeous bouquets to brighten up your home.

What's more, your pollinator garden will feed more than just your honeybees. Just a small portion of your garden dedicated to some of the aforementioned plants can provide a source of pollen and nectar for bumblebees, butterflies, and a wide array of solitary bees and flies (not the kind that fly around trash cans, of course).

USING AN OLD HIVE
TO CATCH SWARMS

While there are quite a few good, inexpensive options on the market for catching swarms, I find that the most sensible way to catch one is to set up a bait hive. This is essentially a small hive with frames of old comb set up and left uninhabited in the spring in order to entice swarms to move in. I love this method because, for one, the hive is set up with components compatible for moving into a bigger hive. You can cobble together a bait hive using frames of comb from an old "dead out," another term for a hive that died over the winter. It'll smell like a beehive, so the swarm's scout bees will find it appealing. To make the hive even more irresistible, put a paper towel with a few drops of lemongrass oil on it within the hive. The lemongrass closely mimics the "come hither" pheromone that bees use to mark a place worth moving into.

If you set up a bait hive in your own apiary, you'll be more likely to catch someone else's swarm. A swarm from your hives will be likely to venture out farther. For this reason, setting up a bait hive is not an appropriate method of swarm prevention.

But be cautious about taking on more bees than you can handle. While bees are fairly low maintenance, during certain times of year, such as swarm season, they can be a little more demanding in terms of management, unless you are unfazed by the prospect of your bees flying off en masse to live someplace else. In urban environments, the issue of competition for slightly more limited resources also comes into play. You only want as many hives as you know can actually sustain themselves in your area. If you live in a very industrial zone, you may want to limit yourself to two to three hives. You don't know who else has bees out there foraging from the same places. And don't forget the native bumblebees and hornets that like to eat some of the same things that bees enjoy!

SWARMING VS. BEARDING

I get panicked texts all the time from new beekeepers in mid-summer. The message will contain an image of a hive with hundreds of bees calmly clustered near the entrance. These uninitiated new-bees (as in the beekeepers) insist that their bees are about to swarm. What they are actually witnessing is an activity known as "bearding." It's a jarring sight the first time you see it, but it's quite harmless. The bees congregate peacefully at their doorstep, occasionally rocking back and forth in a meditative motion beekeepers call "wash-boarding." Bearding usually occurs during the hottest months of the year or when the hive is a little crowded. I liken it to a human experience: The air-conditioning is broken, so the family is just hanging out on the front porch to cool off. In some cases, the hive may be ready for another super. But if that's not the issue, sometimes slightly propping up the outer cover to get the air flowing in the hive will remedy the problem. You can also check to make sure water sources are full so the workers don't have to travel far to get a drink to cool off.

So if your bees start to beard, fret not. A bearding hive is not a swarm. Swarms are loud, active, and a big spectacle. A bearding hive is calm, stationary, and hardly worth getting riled up about.

CHAPTER 7:

Recipes

As a new beekeeper, you will likely view your harvests as a precious substance, best enjoyed in its pure, unadulterated state. The flavors will be distinctive, unlike any honey you'd buy at the supermarket. And the quantity will be limited, making it somewhat rare. You might even be afraid to use your own honey in recipes that might compromise its natural complexity. Why mess with a good thing?

I face the same dilemma each season, but eventually I succumb to the urge to use honey in nearly everything that might otherwise include the flat sweetness of processed sugar. I never regret the decision, as the end result of these substitutions is a more mellow sweetness with notes of flowers, herbs, and spices. In short, honey just makes everything taste better.

The recipes included in this chapter are those that I use on a regular basis because they are easy and satisfying. Like most home cooks, I'm not much for complex processes or long lists of ingredients. I like the beauty of the ingredients to speak for themselves, and, with a food like honey, I believe it is especially important to allow its unique flavor to shine through.

HONEY INFUSIONS: HIBISCUS, VANILLA, AND PINK PEPPERCORN

Makes one 8-oz/240-ml jar

While honey is a pretty stellar treat all on its own, you can have fun experimenting with infusions to amp up its natural complexities. Many herbs and spices will work wonderfully for this recipe, so feel free to try the suggested flavors or get creative with your own combinations. The only rule with infusions is that any ingredients should be dried, as adding anything with any water in it can cause the honey to ferment.

I like to serve these honeys to guests with their tea as an alternative to the "same-old" sugar cubes.

Some of my favorite combinations are white tea with hibiscus-infused honey, oolong tea with pink peppercorn–infused honey, and Earl Grey tea with vanilla-infused honey.

> **1 cup/340 g honey**
>
> **1 tbsp dried aromatic herbs or spices, such as hibiscus flowers, pink peppercorns, rosemary, lavender, or any combination, or 1 vanilla bean, or 1 cinnamon stick**

Fill the bottom of a double boiler with 1 in/2.5 cm of water. Add the honey to the top part of the double boiler and place on medium heat. As the water reaches a low boil, check the temperature of the honey periodically. You'll want to bring it to 185°F/85°C. Once the desired temperature is reached, reduce the heat to low and add the herbs. Simmer for about 5 minutes. Remove the pot from the heat and let stand for 10 minutes. While the mixture is still warm, strain out the herbs and any unsightly particles. When cool, pour the infused honey into sterilized jars, cap, and store in a cool, dark place such as a pantry, for 2 to 4 months.

FRESH GINGER, HERB, AND HONEY ELIXIR

Makes 1 qt/1 L

Ginger and honey are a dynamic duo when combined in any recipe. Both aid digestion and assist our bodies in beating a cold. One of the easiest and most useful recipes in my repertoire is for a fresh ginger and herbal remedy sweetened with a touch of my local honey. A few hot mugs of this stuff will set you right when you're feeling under the weather. I also recommend serving this over ice with a little splash of dark rum at the end of a long day.

> **1 qt/1 L water**
>
> **½ lb/230 g fresh ginger, peeled and grated**
>
> **1 sprig rosemary**
>
> **3 to 4 sprigs fresh mint**
>
> **¼ cup/85 g raw honey**
>
> **Lemon slices, for garnish**

In a stockpot, bring the water to a rolling boil. Add the grated ginger and continue to boil for 20 minutes.

Pour the ginger and water into a 1-qt/1-L heat-safe pitcher or Mason jar. Add the herbs and honey. Cover the jar and let steep for 15 minutes. The ginger and herbs will settle to the bottom, allowing you to pour the elixir into mugs without a strainer.

Serve warm or iced with a slice of lemon.

STRAWBERRY-HONEY LEMONADE

Makes 1 qt/1 L

I remember the first time I sampled this variation of traditional lemonade. I was living in Arizona, and I stopped in a little sandwich shop on a desert-hot midsummer day. A sign in the shop's front window advertising home-style strawberry lemonade made with cactus-flower honey enticed me to venture inside to seek respite from the blazing sun. I downed two of these wonderful drinks in the shop's air-conditioning before braving the heat again. Now whenever summer rolls around, I make sure to stock up on fresh strawberries and lemons so I can whip up a pitcher of Strawberry-Honey Lemonade for my frequent backyard barbecues.

> **1 pt/270 g ripe strawberries, washed, hulled, and halved**
>
> **1½ cups/355 ml fresh lemon juice**
>
> **6 cups/1½ L lukewarm water**

¾ cup/255 g honey

Lemon slices for garnish

Using a blender, puree the strawberries with 1 tbsp of the lemon juice until they are smooth. Set aside.

In a pitcher or 1-qt/2-L Mason jar, combine the remaining lemon juice, water, and honey. Stir until the honey is dissolved. Add the strawberry puree to the lemonade and shake until it becomes pink. If you prefer a less-pulpy lemonade, press the strawberry puree through a sieve before adding it to the lemonade. Pour over a glass of ice and garnish with a slice of lemon. Serve immediately.

SPICED ICED COFFEE WITH HONEY

Makes 6 servings

New Yorkers love their coffee. I think the daily consumption of a cup of joe might even be a requirement to live here. The abundance of local roasters in Brooklyn alone makes it nearly impossible to dodge this addiction.

As you can imagine, an iced-Americano-a-day habit can get expensive. Especially when it's actually a three-iced-Americanos-a-day habit. When times were at their leanest, I started whipping up this delicious version of iced coffee in an attempt to make the standard caffeinated kick-in-the-pants seem pedestrian. This version incorporates aromatic spices like cardamom and cinnamon with a splash of coconut milk and honey, which brings out the fruitier notes in the coffee. (Note: I like to double the amount of coffee grounds for a strong brew that will hold up to the dilution from melting ice cubes.)

1 tsp ground cardamom

1 tsp ground cinnamon, plus 1 cinnamon stick

¾ cup/75 g ground coffee

4 tbsp honey

5 cups/675 g crushed ice

½ cup/120 ml coconut milk, preferably organic

Blend the cardamom, cinnamon, and ground coffee. Brew 6 cups/1.4 L of coffee in the manner you prefer and pour into a heat-safe pitcher. Stir the honey into the brewed coffee until completely dissolved.

Fill six glasses to the top with crushed ice and pour in the sweetened coffee, leaving space at the top of each glass to top off with coconut milk. Garnish with a touch of freshly grated cinnamon stick, stir, and serve!

COUNTRY GIRL JULEP

Makes four 8-oz/240-ml drinks

I love a good cocktail in the summer. And since I usually end up with a bumper crop of fresh herbs from the garden, mixing them into a beverage is a great way to use them and to unwind at the end of a scorching day of rooftop hive inspections. A variation on the classic mint julep, this recipe includes other aromatic, bee-friendly herbs such as basil and rosemary, though you can substitute almost any tender, large-leafed herb.

1 cup/240 ml water

½ cup/30 g fresh mint leaves, plus mint sprigs for garnish

½ cup/30 g sweet basil

½ cup/30 g fresh rosemary

1 cup/340 g honey

1 cup/240 ml good-quality bourbon

Crushed ice

In a medium saucepan, bring the water to a boil. Remove from the heat and add the fresh herbs. Stir until wilted. Add the honey and continue to stir until dissolved. Let the mixture stand until cooled. Strain and discard the mint leaves.

For each julep, combine ¼ cup/60 ml bourbon with 1 tbsp of the honey-mint syrup. Pour the bourbon mixture over crushed ice in each of four frosted tumblers. Garnish with mint sprigs and serve.

THE BEST-EVER HONEY AND OLIVE OIL GRANOLA WITH SEA SALT

Makes about 8 servings

Granola is one of those things I simply cannot live without, and in my opinion, there's no better way to get the day started than with a generous bowl of my own granola, topped with fresh milk. This recipe is my personal favorite of the genre, and it has been tweaked several times over for maximum deliciousness. In short, I like to go all out when I make it, mostly because I know myself well enough to know that if it's completely irresistible, I'll make a point of sitting down to a bowl before I jet out the door in the morning. I swear I hear it calling my name from within the pantry. I suggest adding whatever dried fruits or seeds you love so that yours will do the same for you. (Note: Millet is a highly nutritional African and Asian grain that is now commonly found in health food stores or even large supermarkets.)

> 5 cups/500 g rolled oats
>
> ¼ cup/50 g pearl millet
>
> 4 tbsp/45 g crushed flax seeds
>
> ½ cup/40 g shredded unsweetened coconut
>
> ½ tsp freshly ground cinnamon
>
> ¼ tsp freshly ground nutmeg
>
> ½ tbsp sea salt (the coarser, the better)
>
> ½ cup/120 ml extra-virgin olive oil
>
> ⅔ cup/225 g honey
>
> ½ cup/50 g dried blueberries

Preheat the oven to 375°F/190°C.

Mix the grains, seeds, coconut, and spices together in a large bowl. In a separate medium bowl, mix the olive oil and honey together and whisk until thoroughly incorporated.

Make a reservoir in the center of the oat-millet mixture and pour the honey and olive oil into it. With a large spoon or spatula, fold the oats into the wet ingredients until they are completely coated.

When the grains are completely coated, spread this mixture in an even layer on a large baking sheet and bake in the oven for about 5 minutes, or until the edges of the granola begin to turn golden. Remove from the oven, mix well to ensure even toasting, and return to the oven for another 5 minutes. Repeat this step twice more, making sure the grains are toasting evenly.

Once the granola is consistently golden brown, remove the baking sheet from the oven and let the granola partially cool. Mix in the dried blueberries and transfer to an airtight container, where it will keep for several weeks.

SALTY-SWEET ROASTED NUTS

Makes 8 cups/1.2 kg

For those moments when you need a quick pick-me-up, this snack mix has it all; it's crunchy, salty, sweet, and spicy. Plus it's filling in a healthful way. In other words, it's everything I look for in a snack. I carry a little jar of these tasty nuts with me to help keep me going on a busy day of checking on hives all over New York City. Note that this recipe can easily be adapted for whatever your favorite nuts might be: In my opinion, pecans and almonds are best.

> 2 tbsp olive oil
>
> ¼ cup/85 g honey
>
> 1 tsp cayenne pepper
>
> 8 cups/1.2 kg raw, unroasted nuts
>
> Sea salt

Preheat the oven to 350°F/180°C.

In a large mixing bowl, combine the olive oil, honey, and cayenne. Add the nuts and toss or stir until they are evenly coated with the honey mixture.

Spread the nuts evenly on a baking sheet and put in the oven for 10 minutes. Remove the baking sheet and, using a spatula or wooden spoon, stir the nuts around to ensure they are toasting evenly. Bake for another 10 minutes. Repeat the process of stirring and toasting until the nuts are an even golden brown; this should take about 25 minutes.

Once the nuts are evenly roasted, remove from the oven and, while still warm, season with sea salt. Let cool and store in an airtight container. These will keep for up to 2 weeks.

GRAPEFRUIT SALAD WITH HONEY AND MINT

Makes 6 servings

There is nothing like a light, citrus-studded fruit salad on a hot summer day to help cool you down. This is one of the simplest recipes I've found, and it's a real crowd-pleaser at brunches and barbecues. Whip it up for a last-minute potluck or make it in advance for a lazy weekend breakfast.

> 3 large red grapefruits, peeled and cut into wedges
>
> 4 tangerines, peeled and cut into wedges
>
> 2 tart green apples, such as Granny Smith, washed, cored, and julienned
>
> ¼ cup/15 g mint leaves, minced
>
> ¼ cup/60 ml light spring honey

In a large mixing bowl, mix the grapefruits, tangerines, and green apples. Add the mint. Cover and refrigerate for an hour or until you are ready to serve. Remove the cover, drizzle the honey evenly over the fruit, and serve.

HONEY CROSTINI THREE WAYS

Makes 6 servings

A good loaf of crusty bread is a no-brainer when you're looking for a simple way to serve up local honey. It's great just as it is, although on special occasions I like to spruce up this age-old combination with the addition of cured meats, cheeses, nuts, and fresh herbs. They are perfect for a casual cocktail party or a pre-dinner appetizer, and I prefer using thin slices of comb honey for a little "Wow" factor when putting the crostini together. As an added bonus, each combination makes for a killer sandwich!

2 freshly baked baguettes, sliced into medallions

Olive oil

❀

FRESH RICOTTA, THYME, AND HONEY

1 cup/250 g ricotta

2 tbsp minced fresh thyme

2 tbsp honey

Pinch coarse sea salt

❀

PROSCIUTTO, FIG, AND HONEY

¼ cup/80 g fig jam

8 to 10 slices prosciutto

2 tbsp honey

Pinch of coarse sea salt

1 tsp cracked pepper

❀

GORGONZOLA, BLACK WALNUT, AND HONEY

1 cup/150 g gorgonzola

½ cup/75 g chopped toasted black walnuts

1 sprig rosemary, minced

2 tbsp honey

Pinch of coarse sea salt

Preheat the oven to 350°F/180°C. On a large ungreased baking sheet, arrange the baguette slices, brushing lightly with olive oil. Bake for 5 minutes or until the edges turn golden brown. Let cool.

FOR FRESH RICOTTA, THYME, AND HONEY: Spread a generous layer of ricotta over the toasted bread. Sprinkle with the thyme before drizzling the honey over the crostini. Finish with the sea salt, and serve.

FOR PROSCIUTTO, FIG, AND HONEY: Apply a thin layer of fig jam to the toasted bread. Place a slice of prosciutto on each piece before drizzling with honey. Finish with salt and pepper, and serve.

FOR GORGONZOLA, BLACK WALNUT, AND HONEY: Preheat the broiler before assembling the final crostini. Arrange the toasted bread closely together on a medium baking sheet. In a small bowl, combine the gorgonzola, walnuts, and rosemary. Place a spoonful of the mixture on each slice of bread before placing the baking sheet under the broiler until the cheese mixture is melted and bubbly. Drizzle with the honey, finish with the salt, and serve.

MILK AND HONEY OAT BREAD

Makes 2 loaves

Homemade bread is one of those things I cannot get enough of. Baking this bread in the fall and winter months fills my apartment with warmth and wonderful aromas, and I get the added benefit of enjoying superior sandwiches for the week. This soft milk-bread recipe is really quite easy. It's excellent toasted and then covered with butter and a touch of some rooftop liquid gold.

> **1 tbsp active dry yeast**
>
> **¼ cup/60 ml lukewarm water**
>
> **2 cups/475 ml whole milk, at room temperature**
>
> **1 tbsp honey**
>
> **1 tsp salt**
>
> **2 tbsp unsalted butter, softened**
>
> **1 cup/85 g coarsely ground old-fashioned oats**
>
> **6 cups/900 g bread flour, plus extra for kneading**

In a large bowl or a stand mixer with a dough hook, mix the yeast into the warm water. Add the milk, honey, salt, and butter. Stir until incorporated. Add the oats and 4 cups/600 g of the flour and mix well. Add in enough of the remaining flour to make a sticky dough that follows the spoon around the bowl.

Lightly grease a medium bowl. Turn the dough out onto a lightly floured surface and knead for 10 minutes, adding more flour as needed until the dough is firm and smooth to the touch. Form into a ball.

Put the ball of dough in the greased bowl. Turn the dough over in the bowl, making sure that the top of the dough also gets lightly greased. Cover with a clean cloth and let the dough rise in a warm, draft-free place for 1 hour.

Punch down the dough and turn out onto a lightly floured board. Knead for 5 minutes or until the bubbles are out of the dough.

Grease two 9-by-5-in/23-by-13-cm bread pans. Divide the dough into two equal parts. Shape each half into a loaf. Put each loaf in a prepared pan. Cover and let rise in a warm, draft-free place for 45 minutes or until doubled in size. Meanwhile, preheat the oven to 350°F/180°C.

Once the dough is doubled in size, place the bread pans in the oven and bake for 40 minutes or until the tops are golden brown and the bread sounds hollow when you tap the top.

Remove the bread from the pans, let cool on a baking rack until easily handled, and then enjoy. The bread can be stored in an airtight container for 2 to 3 days.

JALAPEÑO AND HONEY SKILLET CORNBREAD

Makes one 9-in-/23-cm-square loaf

Cornbread really appeals to the Southerner in me. Comforting, sweet, and toothsome, good cornbread is versatile and pairs well with a breakfast of beans and eggs, spicy meat chili, or roast chicken. This recipe gets a little kick from roasted jalapeños, with a touch of velvety honey butter to mellow out the heat.

> 1½ cups/300 g cornmeal
>
> ½ cup/70 g all-purpose flour
>
> 2 tsp baking powder
>
> 1 tsp salt
>
> ½ tsp baking soda
>
> 2 eggs
>
> 1 cup/240 ml buttermilk
>
> 4 tbsp honey
>
> 3 oven-roasted jalapeño peppers, seeded and finely chopped
>
> ¼ cup unsalted butter at room temperature

Preheat the oven to 375°F/190°C. In a bowl, combine the cornmeal, flour, baking powder, salt, and baking soda. In another bowl, whisk together the eggs, buttermilk, and 3 tbsp of the honey. Add the dry ingredients slowly and stir until evenly incorporated. Stir in the jalapeños and then set aside.

Mix 2 tbsp of the butter with the remaining 1 tbsp honey. Place in the refrigerator until ready to use.

On medium heat, melt the remaining butter in a cast-iron or enamelware skillet. Pour the batter in the skillet and smooth out evenly before placing in the oven. Bake for 20 to 22 minutes or until a toothpick inserted near the center comes out clean.

Cut the cornbread into squares or wedges. Serve warm with the honey butter.

SPECK, CHERRY PEPPER, AND HONEY PIZZA

Makes 2 large or 4 individual pies

Is any food more perfect than pizza? Chewy dough, tangy sauce, and a smattering of milky mozzarella never fail to satisfy. In this version, salty cured meat and fiery peppers join forces with a drizzle of honey to produce a pie that starts with a punch and finishes with a kiss.

> FOR THE CRUST
>
> 2¼ tsp active dry yeast
>
> 1 tsp honey
>
> 1 cup/240 ml warm water
>
> 2½ cups/350 g all-purpose flour, plus extra for dusting
>
> 1 tsp salt

1 tbsp extra-virgin olive oil

❊

FOR THE SAUCE

1 cup/255 g San Marzano tomatoes, crushed

2 garlic cloves, minced

2 tbsp good-quality olive oil

❊

¼ lb/115 g thinly sliced Speck Americano, prosciutto, or similar

¾ cup/85 g fresh mozzarella, cubed

½ cup/45 g hot cherry peppers, sliced

2 tbsp honey

TO MAKE THE CRUST: Dissolve the yeast and honey in ¼ cup/60 ml of the warm water in a small bowl. With a mixer fitted with a dough hook, or with a wooden spoon, combine the flour and salt. Add the olive oil, yeast mixture, and remaining ¾ cup/180 ml water. Mix on low speed until the dough comes cleanly away from the sides of the bowl and clusters around the dough hook, about 5 minutes. If the dough is still sticky, add a bit more flour until it pulls cleanly away from the hook.

Lightly grease a large bowl. Turn the dough out onto a clean work surface and knead by hand for 2 or 3 minutes until the dough is smooth and firm. Form the dough into a ball, put in a lightly oiled bowl, and cover with plastic wrap.

Let dough rise for 30 to 45 minutes—until it doesn't rebound when poked with a finger. Divide the dough into either two or four balls, depending on whether you want two large or four small pizzas. Work each ball by pulling down the sides and tucking them under the bottom of the ball. Repeat four or five times. Cover the dough with a damp towel—I use loosely fitted plastic wrap—and let rest 15 to 20 minutes before using.

MEANWHILE, TO MAKE THE SAUCE: Mix together the tomatoes, garlic, and olive oil. Set aside.

Stretch the dough out onto a greased pizza pan, spoon on a thin layer of the sauce, and add toppings to your liking. Bake at 450°F/230°C for 12 to 16 minutes, or until done. For a crispier crust, slip the pizza off the pan directly onto the oven rack for the last 2 minutes of cooking time. Slice and serve.

HONEY BBQ SEITAN SANDWICH WITH QUICK CABBAGE SLAW AND PICKLES

Makes 4 sandwiches

Seitan, a meat alternative made from wheat gluten, does a fine job of sopping up good sauces and making you forget you aren't eating meat. This barbecue recipe uses a South Carolina–style barbecue sauce, and while I won't specify my regional preference for 'cue, I think this style does a great job of complementing the texture and flavor of seitan. And combine sweet, rich barbecue with tart pickles and crunchy slaw, and you'll be making this for every Meatless Monday.

FOR THE BARBECUE SAUCE

1 cup/340 g dark fall honey, or varietals like buckwheat

One 6-oz/170-g can tomato paste

¼ cup/60 ml apple cider vinegar

1 tbsp Dijon mustard

1 tbsp garlic powder

2 tsp freshly ground black pepper

1 tbsp cayenne pepper

1 tbsp hot sauce of your choice

❀

FOR THE CABBAGE SLAW

1½ cups/40 g shredded green cabbage

1½ cups/50 g shredded purple cabbage

1 tart or sweet apple, peeled and julienned

1 small carrot, julienned

½ small red onion, thinly sliced

¼ cup/25 g golden raisins or dried cranberries

4 tbsp light and grassy extra-virgin olive oil, as needed

1 tbsp Champagne vinegar or apple cider vinegar

2 tsp honey

½ tsp caraway seeds

½ tsp freshly chopped dill

Small pinch of ground cumin

Sea salt and freshly ground black pepper

❀

Olive oil

1 small red onion, minced

1 lb/455 g seitan, thinly sliced

4 sesame-seed-topped hamburger buns

Dill pickle slices

Preheat the oven to 425°F/220°C.

TO MAKE THE SAUCE: Combine all the ingredients together in a medium bowl. Let stand at room temperature for 1 hour before using.

TO MAKE THE SLAW: Combine both cabbages, the apple, carrot, sliced onion, and raisins in a large bowl and stir to mix. Drizzle with the olive oil—start with 3 tbsp; you want enough to coat the vegetables but don't want to drown them. Sprinkle the mixture with the vinegar, honey, caraway, dill, and cumin. Season with sea salt and pepper. Toss to coat evenly.

In a lightly oiled, medium sauté pan, fry the minced onion until translucent. Add the seitan and cook, stirring occasionally, until golden brown and slightly crisp on the edges. Add 1 cup/240 ml of the barbecue sauce and simmer for 15 minutes until thickened, stirring occasionally.

Brush the buns with olive oil and toast in a 325°F/160°C oven until golden on the edges. Layer the barbecued seitan on the bun and top with cabbage slaw and dill pickles. Serve immediately!

CRISPY CHICKEN SALAD WITH EGG AND HONEY-MUSTARD VINAIGRETTE

Makes 4 servings

There's nothing like a good salad, especially one that's loaded with all sorts of tasty protein and crispy things. The addition of a homemade honey-mustard dressing helps cut through the richness of the crunchy fried chicken and creamy hard-boiled eggs in this recipe.

FOR THE CRISPY CHICKEN

½ cup/120 ml safflower or vegetable oil for frying

½ cup/70 g all-purpose flour

½ tsp sea salt

1 tsp smoked paprika

3 eggs, beaten

2 boneless, skinless chicken breasts (free-range, if possible), 6 to 8 oz/170 to 230 kg each

½ cup/60 g panko or bread crumbs

❀

FOR THE DRESSING

Juice from 1 lemon

2 tbsp Dijon mustard

2 tbsp honey

½ tsp salt

½ cup/120 ml olive oil

❀

FOR THE SALAD

½ lb/230 g arugula, washed and patted dry

½ cup/80 g thinly sliced red onion

1 whole ripe avocado, cut into ½-in/12-mm cubes

4 hard-boiled eggs, sliced into medallions

TO MAKE THE CHICKEN: Heat a large skillet over medium heat. Add the safflower oil and heat to 375°F/190°C on an instant-read thermometer. On a large plate or in a baking dish, mix the flour with the salt and paprika. Pour the eggs into a shallow bowl. Dust the chicken breasts with the flour mixture, shaking off excess before dipping them into the egg. Roll the chicken in the panko to coat. Add the chicken to the hot oil. Fry the chicken breasts until golden brown, about 5 minutes, flip, and then continue to cook through, about 8 minutes more. Transfer the chicken to a plate lined with paper towels. When cool enough to handle, cut into ½-inch/12-mm strips.

TO MAKE THE DRESSING: Whisk together the lemon juice, mustard, honey, and salt in a small mixing bowl. Slowly add the olive oil, whisking as you pour it in a slow, steady stream until incorporated.

TO MAKE THE SALAD: Toss the arugula, onion, and avocado with the honey-mustard dressing in a large salad bowl. Top with the hard-boiled eggs and strips of crispy chicken. Serve immediately.

HONEY AND SOY-GLAZED DUCK BREAST WITH GARLICKY STRING BEANS

Makes 4 servings

Honey isn't typically thought of as an addition to Chinese-inspired cooking, but it complements the boggy flavor of fermented soy sauce beautifully, adding a hint of sweetness. When combined with fatty, unctuous duck breast, blistered garlicky green beans, and steamed rice, this meal is bound to become a go-to weekday feast.

2 tbsp olive oil

2 fresh duck breasts, skin on

Sea salt and freshly ground black pepper

❀

FOR HONEY AND SOY GLAZE

½ cup/120 ml soy sauce

¼ cup/85 g honey

3 tbsp rice vinegar

1 tbsp black sesame seeds

❀

FOR GARLICKY STRING BEANS

2 to 3 tbsp olive oil (or, duck or pork fat are amazing, if you have either)

3 to 4 large garlic cloves, crushed and coarsely chopped

1 lb/455 g fresh green beans, trimmed, with "strings" removed

Sea salt

In a medium sauté pan over medium-high heat, heat the olive oil. Season the duck breasts with a generous pinch of salt and pepper on both sides. Once the pan is hot, add the duck breasts, searing them until golden brown on both sides. Do not cook them through. Once a nice crust has formed on both sides, move the duck breasts to a plate and cover to retain the heat.

TO MAKE THE GLAZE: Drain off any excess fat from the sauté pan. (You can save it to cook the green beans in!) Over low heat, add the soy sauce, honey, and rice vinegar to deglaze the pan. Stir the mixture in the pan until all of the duck drippings have been incorporated. Continue to simmer and stir for 4 to 5 minutes until reduced by half.

Return the duck breasts to the pan and increase the heat to medium. Cook until the duck breasts are firm to the touch. Remove from the heat and allow to rest for 3 to 4 minutes. Top with a sprinkling of the black sesame seeds.

TO MAKE THE STRING BEANS: In another large sauté pan, heat the olive oil (or duck fat) over medium-high heat. Add the garlic and sauté until lightly golden. Remove the garlic from the pan and set to the side. Add the green beans to the pan and increase the heat to high, stirring the beans frequently. The goal is to give the beans a bit of a char while they cook. Once the beans are bright green and slightly softened, add the garlic back into the pan and continue to cook the beans for an additional couple of minutes, until softened but still vibrant. Remove from the heat. Season the beans with a pinch of salt and serve immediately with the duck breasts.

HONEY AND THYME-GLAZED BBQ PORK RIBS

Makes 2 servings

Grilling is an essential part of life for home cooks here in New York City once the balmy heat of summer arrives. Our tiny apartments heat up fast, making it impossible to use the stove, so those of us with backyards, patios, or park access prefer to cook alfresco.

My favorite marinade for ribs is simple and sweet, and complements the flavor of charred pork beautifully. The addition of honey amps up the rich caramelization on the ribs' edges, while fresh thyme lends a bright, herbaceous touch.

1 lb/455 g pork spareribs, preferably pastured, trimmed of excess fat

Kosher salt and freshly ground black pepper

2 tbsp late-season honey (medium to dark in color)

2 tbsp apple cider vinegar

2 tbsp extra-virgin olive oil

1 large garlic clove, finely chopped

1½ tbsp fresh thyme, minced

In a shallow baking dish, generously season the ribs with salt and pepper. Set aside while you make the marinade.

Whisk together the honey and cider vinegar until totally incorporated. Slowly whisk in the olive oil until it is emulsified. Add the garlic and thyme and whisk well.

Pour the mixture over the seasoned ribs, turning them over so they are evenly coated. Cover and refrigerate for 4 to 6 hours.

Heat the grill to about 400°F/200°C. Place the ribs, meaty-side up, on the grill, away from any hot spots to avoid drying out the meat. It is not necessary to flip them. The ribs should be cooked through in 25 to 30 minutes, depending on their thickness. Cut ribs individually and serve immediately.

MOLASSES-FREE GINGERBREAD COOKIES

Makes 24 cookies

Nothing quite warms the soul and conjures the spirit of winter like spiced cookies. I just love them. I make batches of these for friends and family for the holidays but am always sure to make twice as many as I need because my boyfriend and I end up eating half of them.

> 2¼ cups/315 g all-purpose flour
>
> 2 tsp ground ginger
>
> 1 tsp baking soda
>
> ¾ tsp ground cinnamon
>
> 2 tbsp cocoa powder
>
> ½ tsp ground cloves
>
> ¼ tsp salt
>
> ¾ cup/175 g butter, softened
>
> 1 cup/340 g dark honey
>
> 1 egg
>
> 1 tbsp water

Preheat the oven to 350°F/180°C.

Sift together the flour, ginger, baking soda, cinnamon, cocoa powder, ground cloves, and salt. Set aside.

In a large bowl, cream together the butter and honey until incorporated. Beat in the egg and then stir in the water. Gradually stir the sifted ingredients into the honey mixture.

Shape the dough into walnut-size balls, placing the cookies 2 in/5 cm apart on an ungreased baking sheet. Flatten each ball slightly with the bottom of a glass. Bake for 8 to 10 minutes, then let cool on the baking sheet for 5 minutes before removing to a wire rack to cool completely. Store in an airtight container. These will keep for up to 1 week.

HONEY "TREACLE" TART

Makes one 9-in/23-cm tart

I've always loved English-style desserts, especially the richly sweet treacle-based ones. For those who are uninitiated, treacle is a syrup not unlike molasses. It's a product of the sugar refining process and comes in light and dark varieties. I always feel a bit guilty using it when I've got honey, which is arguably superior, in my pantry. The use of local honey in this traditional English treat adds a nice mellow complexity to the final result, which is usually so sweet it makes your teeth ache.

> **FOR THE PASTRY**
>
> 1¼ cups/175 g all-purpose flour
>
> Pinch of fine sea salt
>
> ¼ lb/114 g unsalted butter, chilled
>
> 1 egg, beaten
>
> 2 or 3 tbsp ice water

❀

FOR THE FILLING

½ cup/50 g old-fashioned oats

¼ tsp ground ginger

1 cup/340 g honey

1 tsp lemon zest, plus 2 tbsp lemon juice

TO MAKE THE PASTRY: In a large mixing bowl, combine the flour and salt. Mix well and then rub in the butter by hand until the mixture resembles coarse meal with pea-size beads of butter throughout.

In a small bowl, combine the egg and 2 tbsp of the ice water. Whisk together and then gradually add to the flour mixture, stirring with a fork. Mix until the dough forms a ball. Add the remaining 1 tbsp water if necessary.

Press the dough into a smooth ball, cover with parchment paper, and allow to rest in the refrigerator for half an hour.

Preheat the oven to 400°F/205°C.

Roll out the dough into a 9-in/23-cm tart pan. Trim the edges and reserve the scraps.

TO MAKE THE FILLING: In a medium mixing bowl, incorporate the oats, ginger, and lemon zest with the honey and lemon juice before transferring the mixture to the prepped tart pan. Pat the filling down gently with your fingertips.

Roll out any remaining pastry and cut into strips to lay a lattice over the tart, if desired. Bake for 30 minutes until the crust is golden brown. Allow the tart to cool for 10 to 15 minutes before cutting into it. Best served warm.

GOAT'S MILK ICE CREAM WITH CARDAMOM, DATES, AND HONEY

Makes 12 cups/2¾ L

I don't get to eat ice cream very often, but when I do, it's often stuff I make at home. This recipe is excellent for those who have issues with lactose and prefer to avoid sugar. It's sweet, light, and wonderfully aromatic. Bonus if you have your own goats for the milk. The fresher the milk is, the better and more mild the flavor. This recipe requires an electric or hand-crank ice-cream machine. Serve with bits of comb honey.

8 cups/1¾ L goat's milk

1 cup/340 g honey

½ tsp ground cardamom

6 egg yolks

1 tsp vanilla extract

1 cup/75 g chopped dates

1 cup/120 g chopped pistachios

In a medium saucepan, combine the goat's milk, honey, and ground cardamom. Bring to a simmer over medium heat.

In a separate medium saucepan, whisk the egg yolks until smooth. Remove the milk mixture from the heat and let cool for 5 minutes. Skim off any skin that forms on the surface.

Using a ladle, slowly pour the warm milk mixture into the yolks while continually whisking to prevent the yolks from cooking. Once all the milk is added, add the vanilla, then pour the mixture into the ice-cream maker and set the machine using the recommended factory settings. This recipe should take about 25 minutes or so to come together. Once a thick, custardy consistency is reached, spoon the ice cream into a frozen glass mixing bowl. Add the dates and pistachios, folding them into the ice cream. Pack into a freezer-safe lidded container and store in the freezer for up to 3 weeks.

CLOVE AND HONEY ORANGE MARMALADE

Makes about six 8-oz/240-ml jars

Marmalade is one of those foods that people assume is difficult to make, but I can attest to the simplicity of this bittersweet preserve. Unlike many other jams and jellies, you don't need to use store-bought pectin, as it's already contained in abundance in the peels of the citrus fruit. All that is needed is a little bit of time and a touch of patience. The addition of sugar to this recipe actually helps with the set of the jam. Feel free to reduce or omit the sugar if you don't mind a slightly softer marmalade.

4 large seedless oranges, such as tangelos

2 lemons

8 cloves

1 stick of cinnamon

8 cups/1¾ L water

2 cups/400 g sugar

2 cups/680 g honey

Cut the oranges and lemons in half crosswise and then cut the halves into very thin half-moon slices. (If you have a mandoline, this will be easy.) Discard any seeds. Put the sliced fruit and their juices into a stainless-steel pot with the cloves and cinnamon inside a muslin sachet. Add the water and bring the mixture to a boil, stirring often.

Remove the mixture from the heat and stir in the sugar and honey, until the sugar dissolves. Cover and let stand overnight at room temperature.

The next day, remove the spice sachet and bring the mixture back to a boil. Reduce the heat to low and simmer, uncovered, for about 2 hours. Reduce the heat to medium and boil gently, stirring often, for another 30 minutes. Skim off any foam that forms on the top.

Continue to cook the marmalade until it reaches 220°F/105°C on a candy thermometer. If you want to be doubly sure it's ready, place a small plate in the freezer for 20 minutes. When marmalade reaches temperature, test the set by removing the plate from the freezer and spooning a bit of the marmalade onto it. If it gels upon contact with the plate—not runny but not hard—you've got a suitable set.

Pour the marmalade into clean, hot

canning jars. Wipe the rims thoroughly with a clean damp paper towel and seal the jars with rings and lids. To shelf stabilize, place the lidded jars in a large pot of boiling water for 15 minutes. Carefully remove and invert the jars on a clean, dry towel. Once cool enough to handle, flip the jars over and test the seal by pressing on the top of the lid. If the lid makes a popping sound, place back in the boiling water for another 10 minutes. If there is no pop, you've achieved a proper seal.

Jars of marmalade can be stored at room temperature in the pantry for up to 1 year. Once opened, jars must be refrigerated and will last 2 to 3 months.

HONEYCOMB CANDY WITH ROSEWATER

Makes about 36 pieces

Holidays are a stressful time for many of us. If you are anything like me, your friends and family have more doodads and thingamajigs than they know what to do with. Giving handmade treats during the holidays is my favorite solution to the gift-giving dilemma.

This recipe appeals to my curmudgeonly side and is great for last-minute gifting. It is also a frugal way to share the fruits of your bees' labor. Note that many people will dip these in melted chocolate, but I prefer them just like this, as you can taste the honey, and the rosewater adds even more floral notes to the candy.

1 cup/200 g sugar

2 tbsp light- to medium-grade honey

2 tbsp rosewater

1¼ tsp sifted baking soda

Grease a baking sheet or place a Silpat mat on your kitchen counter.

In a heavy, medium saucepan, add the sugar, honey, and rosewater. Mix until thoroughly incorporated and then continue melting the mixture over high heat.

Once the liquid is bubbling, lower the heat to medium. Cook for about 5 minutes or until golden, stirring occasionally.

Remove the mixture from the heat and immediately add the baking soda, mixing it in quickly. The mixture will expand and foam. As it does, quickly pour it onto the prepared baking sheet. Allow the mixture to cool for 20 minutes. Then, with a serrated knife, cut or break the candy into bite-size pieces or bars.

Once the candy is completely cool, store in an airtight container in a cool place up to 1 week.

HEALING HONEY SALVE

Makes one 8-oz/240-ml jar

This is one of my favorite everyday beauty products. It has so many uses beyond smoothing the skin of overworked hands. I also use it as a balm for my hair, as a cuticle cream, and as a foot balm. I've even been known to put it on my lips during the dry cold of winter.

¼ cup/40 g grated beeswax

1 cup/240 ml olive oil

10 drops lavender oil

In a double boiler, completely melt the beeswax. Using a spatula, gradually mix in the olive oil and lavender oil until well combined.

Pour the mixture into a clean, dry 8-oz/240-ml glass canning jar to cool. Seal the jar and let set for 48 hours before using.

PROPOLIS TINCTURE

Makes one 8-oz/240-ml jar

Propolis, the resinous substance bees use to sanitize and strengthen their hive, has long been reported to have tremendous antibacterial, antifungal, and antimicrobial properties. Often referred to as "bee glue," propolis has been used topically to heal fungal infections, canker sores, and wounds and has also been used for centuries as a remedy for the common cold, among many other ailments. This tincture recipe is very easy to make and can be added to hot tea for a daily supplement. The recommended dosage is six to ten drops a day, or about ¼ tsp, but be sure to consult your physician before taking this tincture, as it is powerful and could have an adverse effect on those with bee-related allergies.

¼ cup/35 g propolis

1 cup/240 ml 100 proof vodka, grain alcohol, or whiskey

Put the propolis in the freezer for 1 or 2 hours until the normally gummy substance is brittle. Place the frozen propolis between two sheets of wax paper and crush with the bottom of a heavy saucepan, working quickly, until the propolis is broken into finer pieces. Pour them into a glass jar.

Pour the vodka over the propolis, cover the jar, and give it a good shake. Let the tincture sit for about 2 weeks, shaking daily. At the end of the 2 weeks, strain the liquid through cheesecloth and pour into a clean bottle.

Store the propolis tincture in a cool, dark place, such as a medicine cabinet, where it will keep for up to 1 year at room temperature.

HONEY RHASSOUL CLAY FACIAL MASK

Makes a single treatment

Rhassoul (also known as *ghassoul*) is a soft, mineral-rich clay harvested from the mountains of Morocco. It's been used for ages as a hair and facial cleanser and has been proven to reduce skin dryness and improve skin clarity. With the addition of the humectant properties of honey, this mask performs miracles for your complexion during the coldest, most arid months of the year and also works impressive magic as a summer skin detoxifier.

2 tbsp Moroccan rhassoul clay (available at MountainRose Herbs.com, a personal favorite)

2 tbsp raw honey

Cool water as needed

Put the clay in a small bowl. Add the honey ½ tbsp at a time, incorporating each addition before adding more. Once all the honey has been incorporated into the clay, add a few splashes of water to thin the mixture to the consistency of toothpaste.

To use, apply evenly to your face, avoiding the eye area. Find a quiet place to lie down, and drape a warm, moist towel over your face for up to half an hour. Before removing the mask, rub it over your skin in circular motions to get the added benefit of exfoliation. Rinse your face with lukewarm water and gently pat it dry.

BEEHIVE STAIN-N-SEAL

Makes 2 qt/2 L

Not everything that comes from the hive needs to be eaten to be useful. We humans have long used propolis, one of my favorite bee-related substances, the same way the bees use it inside of their colony—as a protectant and sealant. This is an easy and inexpensive way to finish the exterior of new hives or furniture, and it imparts a wonderfully warm color to pale woods like pine and cedar.

1 qt/1 L blond shellac

6 oz/170 g raw propolis

1 qt/1 L denatured alcohol

Combine all ingredients in a glass jar at room temperature. Cover the jar with a lid and allow the mixture to stand at room temperature for 2 weeks, gently shaking and rotating it at regular intervals.

At the end of the 2 weeks, filter the solution through a few layers of cheesecloth or a nylon stocking before using.

Before applying the varnish, sand the surface of the wood and wipe it down with a clean dry cloth. Brush on two or three coats of the shellac, layering evenly with a fine-bristle paintbrush and allowing the surface to completely dry overnight between coats.

EASY BEESWAX CANDLES

Makes 12 candles

Beeswax is ideal for making candles thanks to both its naturally aromatic qualities and its clean burning habit. As a beekeeper, you'll likely end up with quite a bit of wax from your honey harvests, and this easy candle recipe will help ensure you'll never need to buy candles again.

MATERIALS YOU WILL NEED:

Cotton or hemp wick

Wick tabs

1 case 4-oz/120-ml canning jars with lids and rings

2 lb/910 g beeswax, grated

2 to 3 drops of your favorite essential oil (optional)

1 lb/455 g soy wax beads, which can be purchased from any local craft store

❀

TOOLS YOU WILL NEED:

Pliers

Superglue

Cut your wick into lengths of about 6 in/ 15 cm and feed them through the wick tabs. Use your pliers to close the open end of the tab so the wick will not fall out.

Superglue a wick-and-tab combination to the bottom of each jar, centering each. Make sure to position the wick upright. Fill the jars to the brim with two parts grated beeswax to one part soy wax beads. Add the essential oil, if you wish. Trim the wick to the same height as the edge of the jar. Repeat the process until all of the jars are filled. Meanwhile, preheat the oven to 200°F/95°C.

When the candles are prepped, put them on a baking sheet and place in the oven. Watch for the wax to melt completely—about 20 minutes.

Once the wax has melted, pull the baking sheet out of the oven very carefully. Let the candles cool and the wax reharden. Trim your wicks again to about ½ in/ 12 mm from the wax surface.

After the candles are completely cool, put the lids back on the jars before storing in a cool dark place, or giving away.

GLOSSARY

ABSCONDING: a phenomenon in which a bee colony completely abandons the hive, typically during a broodless period, taking some or all of their food stores. Causes of this behavior can be frequent disruption, lack of food, and abnormal weather patterns. This is not to be confused with swarming, in which the colony divides, leaving a young or developing queen, brood, and nearly half of the workforce.

ACARINE DISEASE: a less frequently used term for tracheal mites.

AFRICANIZED HONEYBEE: known as the "killer bee" in popular culture, these hybridized crosses between the African honeybee and the European honeybee are often classified by their extreme defensiveness and resistance to disease.

APIARY: a specific location in which one or several hives are kept.

APICULTURE: the study and practice of beekeeping.

APIS CERANA: the Latin name for the Asiatic honeybee.

APIS MELLIFERA: the Latin name for the European honeybee.

APITHERAPY: the use of bee venom, wax, propolis, honey, and royal jelly to promote healing in the human body.

BEE BREAD: a fermented mixture of pollen, nectar, and glandular secretions stored in cells to be fed to larva and young bees.

BEE BRUSH: a soft-bristled brush used for gently removing bees from areas where a beekeeper would like to work.

BEE SPACE: a popularly accepted measurement theorized by Lorenzo Langstroth that supposes that spaces larger than ⅜ in/10 mm will inspire the bees to build brace comb, and spaces smaller than this will inspire the application of propolis by bees. Langstroth hives are constructed with this space in mind.

BEE YARD: a specific location where several hives are maintained.

BOTTOM BOARD: the bottom-most level of a Langstroth hive. Screened bottom boards, which can be bought or made, are recommended for rooftop apiaries because they help increase airflow and allow the bees to manage mite infestations more effectively on their own.

BRACE COMB (ALSO, BURR COMB): errant comb built in between or on top of frames, which can make inspection more challenging.

BROOD: a general term for eggs, larvae, and pupae.

BROOD CHAMBER (OR BROOD NEST): the part of the hive dedicated to brood production. It is typically located in the lower half of most modern hives but can vary in feral hives.

CAPPED BROOD: cells of developing bees covered with a thin layer of wax and propolis. Worker bees seal the larvae into these cells, where they become pupae and, eventually, adult insects.

CASTE: a system in honeybee colonies in which bees in various stages of adulthood perform specific duties (i.e., house bees, guard bees, forager bees).

CELL: the wax structure in a hive in which food is stored and individual bees are grown.

CHILLED BROOD: a condition in which brood dies due to insufficient warmth.

CLEANSING FLIGHT: close-range flights that honeybees will take in order to defecate. Bees will not do so inside of the hive unless they are ill.

CLUSTER: a tightly packed ball of bees, varying in size. Bees instinctively cluster when swarming to stay intact and protect the queen or to retain warmth during the winter months.

COLONY: a family of bees contained in a hive that includes a queen, workers, drones, and brood.

COLONY COLLAPSE DISORDER (CCD): a complex and yet unknown malady affecting large numbers of colonies, especially within the commercial and migratory beekeeping industry. It is suspected that a combination of pathogens, stress, and insufficient nutrition are to blame.

COMB: a wax structure built by the honeybee in early adulthood in which to rear brood and store pollen and nectar.

CRYSTALLIZATION: the formation of sugar crystals in honey. Crystallized honey is edible but has increased in viscosity to the point of no longer being pourable. Crystallized honey can be returned to its original liquid state through heating.

DEAD-OUT: a beekeeping term for a hive that has died over the winter.

DRIFTING: the unchecked passing of drones from one colony to another.

DRONE: a male honeybee.

EGG: the primary stage in the metamorphosis of the honeybee.

ENTRANCE REDUCER: any item used to plug up a portion of the hive entrance to allow a small colony to defend itself more easily. Beekeeping suppliers sell a variety of simple devices for this purpose.

EXTRACTOR: a mechanical device (also known as a centrifuge) used for removing honey from the comb without damaging the comb. This allows beekeepers to give comb back to the bees to reuse for several seasons.

FEEDER: a container used to provide bees with sugar syrup to supplement their food intake.

FERAL COLONY: a honeybee colony that occupies a space of its own choosing and is not maintained by a beekeeper.

FOOD MILES: the distance food travels to get from farm to consumer.

FORAGER: the final stage in a honeybee's adult life. Forager bees are responsible for finding and retrieving nectar, pollen, water, and propolis.

FOULBROOD: a bacterial brood disease, typically indicated by ropey, "snot-like" brood and an unpleasant smell coming from the hive. There are two commonly known strains of foulbrood: American and European. American foulbrood is the more virulent, and hives bearing symptoms are often required by law to be destroyed by burning them.

FOUNDATION: a sheet of pressed beeswax embossed with a cell structure for building wax combs that the bees follow in creating comb.

FOUNDATIONLESS: a method of beekeeping in which bees are allowed to build wax without the aid of pressed foundation.

FRAME: a rectangular wooden or plastic structure present in Langstroth hives intended to hold comb, making it easy to remove during inspections.

GREASE PATTY: a mixture of sugar, shortening, and usually wintergreen oil used in the treatment of Varroa and tracheal mites.

GUMS: an old form of beehive, bee gums consist of a hollowed-out log in which the bees are allowed to build natural comb. These hives are nearly impossible to inspect and therefore illegal in many countries.

HIVE BODY: this term is interchange-able with "brood chamber" but refers more specifically to the lower portion of the hive containing the brood nest.

HIVE STAND: a simple structure upon which a hive is situated.

HIVE TOOL: a small steel implement similar in shape to a crowbar, used for prying apart and removing frames from a hive.

HONEY FLOW: refers to the rate of nectar production in the local flora.

HONEYCOMB: celled structure within the hive that is made of wax and contains cured honey.

HOUSE BEES: young worker bees that occupy the hive during the first few weeks of adulthood, before graduating to foragers.

HYGIENIC BEHAVIOR: preening and polishing of brood and wax demonstra-ted by house bees. Hygienic behavior is desirable and helps colonies cope with pest and disease pressure.

INNER COVER: a board placed upon the uppermost hive body or super, underneath the outer cover. Most inner covers are notched to help increase ventilation and to serve as a secondary entrance for foragers.

INTEGRATED PEST MANAGEMENT (IPM): a method of hive management that relies on the least-invasive or least-disruptive measures to help ensure colony health.

K-WING: a symptom of several honeybee maladies such as tracheal mites or Nosema. It can be observed when the humuli that hold the fore and hind wings of the bee together come undone, causing the wings to splay. They look like the letter "K" as they protrude from the bee's thorax.

LANGSTROTH HIVE: a style of hive developed in 1852 by Lorenzo Lorraine Langstroth. It contains movable frames and hive components for easy manipula-tion and remains the most common type of hive used by beekeepers today.

LARVA: the second stage of honeybee development. Occurs from day four to day nine or ten of a bee's life.

LAYING WORKER: a potentially serious malady that occurs when there has been little or no queen pheromones being distributed in a colony for a length of time. Young house bees begin laying unfertilized eggs, which become drones.

MARKED QUEEN: a queen, usually produced commercially, that has been marked on her thorax with a color-coded dot that indicates the season she was reared and allows for easier location during inspections.

NECTAR: a sweet liquid that bees collect, produced by flowering plants.

NOSEMA DISEASE: also called *Nosema apis,* this is a fungal disease that causes dysentery-like symptoms. Flare-up usually occurs during the winter when bees are confined and cannot go on regular cleansing flights.

NUCLEUS HIVE (OR "NUC"): a miniature hive that consists of three to five frames of brood, stored food, worker bees, and a queen.

NUPTIAL FLIGHT (OR "MATING FLIGHT"): a series of flights that a virgin queen takes to mate with several drones in the area. Once mated, she returns to the hive to begin laying eggs.

NURSE BEE: one of the many hats worn by young adult worker bees. The bee takes on this role shortly after emergence. Nurse bees are responsible for feeding and grooming larvae and tidying up the brood nest.

"OPEN-MATED" QUEEN: a queen bee that is allowed to breed naturally with whatever drone stock is in the immediate area.

OUTER COVER: the top component of a beehive. It sits above the inner cover and serves as a protective shield from the elements.

PACKAGE: a box containing approximately 3 lb/1.4 kg of bees with a caged queen and a can of syrup feed for transport. Package bees are one of the most common ways of procuring bees for a new hive.

PHEROMONE: a chemical excreted by all bees that communicates need or function within the colony.

POLLEN: a high-protein substance produced by male flowers. Pollen is a vehicle for transporting genetic information between plants in the process known as pollination.

POLLEN PATTY: a formed cake of pollen or pollen substitute mixed with syrup and pressed between wax paper. Placed between hive bodies, this is an excellent way to help induce brood rearing.

PROPOLIS: a mixture of tree resin and small amounts of wax used as a structural stabilizer, a sanitizing agent, and a sealant by honeybees.

PUPA: the third stage in the metamorphosis of a honeybee. At this stage, the cell containing the bee is capped over until the pupa emerges as an adult. This final stage takes between seven and fifteen days depending on the gender of the bee.

QUEEN: the matriarch of a colony, charged with the task of reproduction. She is typically the only sexually mature female bee in the hive.

QUEEN CELL: a large, peanut-shaped cell containing a developing queen.

QUEEN CUP: the beginning stages of a queen cell. Most colonies will produce several queen cells in the brood nest as a form of insurance against queenlessness.

QUEEN EXCLUDER: a metal or plastic grate placed over the brood nest to keep the queen from laying eggs in honey supers. This is also known as a "honey excluder," as some workers do not like passing through an excluder, so they store more honey in the brood nest.

QUEENRIGHT: a term used to describe a colony with a viable, laying queen.

QUEEN SUBSTANCE: the pheromone exuded by the queen bee.

REQUEEN: to replace a failed queen or one with undesirable traits with a queen of the beekeeper's choosing.

REVERSING (AS IN "REVERSING HIVE BODIES"): a second-season management technique in which the two hive bodies are swapped to give the colony more room to rear brood.

ROBBING: an occurrence during which bees from foreign colonies invade weakened hives to steal food. This usually occurs during a nectar dearth.

ROYAL JELLY: a nutritionally dense substance secreted by nurse bees and fed to young brood. It is fed in more generous amounts to developing queens, speeding their development and ensuring they will be reproductively viable.

SCOUT BEE: a bee tasked with the specific duty of finding a potential new home previous to swarming, as well as locating abundant food sources.

SKEP: an old-world basket-type hive made from woven reeds. Skeps are illegal in the United States, as they do not have the movable combs required to perform health checks.

SMALL CELL: a type of foundation used by many beekeepers to increase a colony's ability to better manage maladies and pests. Small-cell foundation typically has a cell imprint of 4.9 mm, where standard-cell foundation has a cell size of 5.4 mm. Foundation with a cell size of 5.1 mm is available for gradual regression to small cell through Dadant.com.

SMOKER: an essential tool composed of bellows that force air through a spouted combustion chamber. Flammable organic matter is burned in the chamber and then snuffed so that it smolders, producing a cool smoke that, when puffed into the hive, disrupts the flow of pheromones that could put honeybees on the defensive.

SUPERCEDURE: the process a colony goes through to replace a failing queen.

SUPERING: a term describing the addition of boxes and frames designated for honey storage to a hive. Honey supers are placed above the brood nest.

SWARM: the reproductive division of a robust colony that has outgrown its current home. When a colony swarms, approximately 60 percent of the workforce absconds with the current queen, leaving behind a population of house bees, brood, stored food, and several queen cells close to emergence.

SWARM TRAP: a hivelike device set up to appeal to and catch swarms.

TOP-BAR HIVE: a style of hive consisting of a vessel, usually a rectangular box, with bars placed along the top for the bees to build natural comb onto.

TRACHEAL MITE: a small mite that feeds and reproduces in the trachea of young bees. This type of mite cannot be seen with the naked eye but is often indicated by the presence of K-wing and weak and disoriented bees found crawling on the ground in front of the hive. To positively diagnose tracheal mites, a specimen collection must be sent off to your local state agricultural extension laboratory to confirm the presence of these parasites.

TREATMENT-FREE: a method of beekeeping that avoids all treatment of honeybee maladies and instead encourages small or natural cell regression, genetic diversity, and selective breeding.

VARROA MITE: *Varroa destructor*, the most common and aggressive parasitic pest to *Apis mellifera*. The Varroa mite feeds on the hemolymph, or blood, of the bee and serves as a vector for other pathogens.

VENOM: a toxin administered through the stinger of the worker bee that increases the effectiveness of stinging.

WALK-AWAY SPLIT: the process of creating multiple hives from one crowded, usually queenright colony through the process of splitting up frames of uncapped and open brood, adult bees, and stored food.

WAX MOTH: a destructive insect invader that lives off of brood comb.

WINTERIZING: the process of preparing a hive for the cold winter months, when the colony is dormant, though still very much alive.

WOODENWARE: a general term for any of the wooden components of a beehive. This includes frames, supers, and hive bodies.

WORKER BEE: a sexually immature female bee, responsible for all of the work performed. A normal colony is made up of about 85 percent workers.

RESOURCES

Beekeeping Supplies

GENERAL BEEKEEPING:
Betterbee, Inc.
8 Meader Road
Greenwich, NY 12834
Phone: 1-800-632-3379
www.betterbee.com

Brushy Mountain Bee Farm
610 Bethany Church Road
Moravian Falls, NC 28654
Phone: 1-800-BEESWAX
www.brushymountainbeefarm.com

Dadant & Sons Inc.
51 South 2nd Street
Hamilton, Illinois 62341
Phone: 1-888-922-1293
www.dadant.com

Mann Lake, Ltd.
501 1st Street South
Hackensack, MN 56452
Phone: 1-800-880-7694
www.mannlakeltd.com

Walter T. Kelley Co.
PO Box 240
807 W. Main Street
Clarkson, KY 42726
Phone: 1-800-233-2899
www.kelleybees.com

Specialty Hives

LANGSTROTH WOODENWARE:
Evans Cedar Beehives
PO Box 270
Marmora, NJ 08223
Phone: 1-856-457-4572
www.evanscedarbeehives.com

TOP-BAR HIVES, KITS AND PRE-MADE:
Bee Thinking
1551 SE Poplar Avenue
Portland, OR 97214
Phone: 1-877-325-2221
www.beethinking.com

Gold Star Honeybees
PO Box 1061
Bath, ME 04530
Phone: 1-207-449-1121
www.goldstarhoneybees.com

Honey and Disease Testing

USDA Agricultural Service
Bee Research Laboratory
Bldg 476, Room 204
Beltsville Agricultural Research
Center-East
Beltsville, MD 20705
Phone: 1-301-504-8821
www.ars.usda.gov

Recommended Reading

The ABC and XYZ of Bee Culture
By A. I. Root, E. R. Root
A.I. Root Company, 2007

The Backyard Beekeeper: An Absolute Beginner's Guide to Keeping Bees in Your Yard and Garden
By Kim Flottum
Quarry Books, 2010

Beekeeping at Buckfast Abbey
By Brother Adam
Northern Bee Books, 1987

The Hive and the Honeybee
By Roy A. Grout
Dadant & Sons, 1992

Honeybee Democracy
By Thomas D. Seeley
Princeton University Press, 2010

The Practical Beekeeper; Beekeeping Naturally
By Michael Bush
X-Star Publishing, 2011

Top-Bar Beekeeping
By Les Crowder
Chelsea Green Publishing, 2012

Recommended Websites

American Apitherapy Society
Great resource for people interested in or practicing apitherapy.
www.apitherapy.org

American Beekeeping Federation
Information about local beekeeping conferences, scientific studies, and legislation. Memberships available.
www.abfnet.org

Anarchy Apiaries
Treatment-free beekeeping, top-bar hives, and queen breeding from rambling beekeeper Sam Comfort.
www.anarchyapiaries.org

Apimondia
The world's largest annual international beekeeping conference.
www.apimondia.org

Backwards Beekeepers
Sustainable urban beekeepers and swarm rescue in Los Angeles, California.
www.backwardsbeekeepers.com

Beesource Beekeeping
Forums, chat, and essays written by knowledgeable beekeepers.
www.beesource.com

Brooklyn Grange
New York City commercial apiary and urban beekeeping apprenticeship program.
www.brooklyngrangefarm.com

Learning Beekeeping
How-to videos and essays focused on alternative hive design and management practices.
www.learningbeekeeping.com

Pollinator Partnership
Resources for pollinator-friendly gardening, eco-activism, and education.
www.pollinator.org

INDEX

ACKNOWLEDGMENTS

This book materialized, in no small part, due to the inspiration of other urban beekeepers. Some friends, some rivals, some respected mentors, all had a part to play in the writing of *The Rooftop Beekeeper*. I owe much of what I know or have forced myself to learn to their insight, strong personalities, and vastly differing opinions. I'd like to thank Kirk Anderson and Sam Comfort for their authentic spirit and for teaching me the importance of "just letting them bee." This book would not be what it is without the support of Kelly York and Timothy O'Neal who, above many other things, were perfect model beekeepers during the bee-handling photo shoots. Big thanks to my friends Chase Emmons from Brooklyn Grange and Michael Leung of HK Honey, who are constantly going out of their way to prove the viability of urban agriculture as a sustainable business model. There are too many more to mention, but all of you are huge sources of inspiration to me and I feel blessed to know you.

Many of my first apiary sites and teaching opportunities came about because of help from New York City's urban farmers. I'd like to thank the folks from Brooklyn Grange, Annie Novak at Eagle Street Rooftop Farm, and Stacey Murphy of BK Farmyards for giving me an outlet to geek out about bees in front of other aspiring apiarists.

I'd like to give extra-special thanks to the women in my family, especially my mother Wanda. She, in her own way, taught me the importance of resilience and fearlessness. Without those qualities I don't think I would have made a home in Brooklyn or stayed a beekeeper for very long. And to my Grandmother Phyllis and Great-Grandmother Myra, whose occasional brassiness, softened with genteel Virginian humor, always demonstrated the kind of character to aspire to. Without these women in my life, I may have given up as I hit the inevitable bumps in the road as a beekeeper and first-time author.

I'd like to express my gratitude to the folks at Chronicle Books, especially Bill LeBlond and Sarah Billingsley, for taking a chance on a rough-and-tumble kid from Baltimore who likes to play with big boxes of bees. Enormous thanks to Alex Brown for going above and beyond as a photographer and a friend. Also, to Rachel Wharton for all of her encouragement, professionalism, and good nature. Thank you to Masako Kubo for elevating it all with her beautiful illustrations. I feel so honored to have my name listed among such talent.

Last but not least, thank you to Neil Despres, my dearest friend, for never making me feel crazy for wanting to be a rooftop beekeeper.